THOMISTIC EVOLUTION

OD SAW THAT IT WAS GOOD.

IN THE IMAGE OF GOD HE CREATED THEM.

THROUGH

IN THE

HIM ALL THINGS WERE MADE.

BEGINNING

Thomistic Evolution

*A Catholic Approach to
Understanding Evolution in the Light of Faith*

SECOND EDITION

Nicanor Pier Giorgio Austriaco, O.P.
James Brent, O.P.
Thomas Davenport, O.P.
John Baptist Ku, O.P.

Dominican Friars of the Province of St. Joseph

CLUNY MEDIA

Cluny Media second edition, 2019

For more information regarding this title
or any other Cluny Media publication,
please write to info@clunymedia.com, or to
Cluny Media, P.O. Box 1664, Providence, RI 02901

VISIT CLUNY MEDIA ONLINE AT
WWW.CLUNYMEDIA.COM

VISIT THOMISTIC EVOLUTION ONLINE AT
WWW.THOMISTICEVOLUTION.ORG

ISBN: 978-1950970797

Cover design by Clarke & Clarke
Cover image: Based on Ernst Haeckel, "Plate 2: *Thalamphora*,"
from *Kunstformen der Natur* (1900)
Courtesy of Wikimedia Commons

CONTENTS

Preface to the Second Edition

Since the publication of the first edition of *Thomistic Evolution*, Fr. Austriaco has revised his account of the evolution of our species and now defends the historicity of an original human being, the human being named Adam. The chapters on the historicity of Adam and Eve have been rewritten to reflect this development. Moreover, a new chapter has been added to respond to criticisms of Thomistic evolution made since the publication of the first edition of this book.

❧ INTRODUCTION

Evolutionary theory, understood here as the scientific claims that all the living organisms on our planet have a common biological origin and that these diverse organisms arose through a process of natural selection acting on genetic diversity, has raised numerous disputed questions among the Catholic faithful and other Christian believers. Some of these include the following: How does God create in an evolving world? How are we to read and interpret the creation narratives found throughout the sacred Scriptures in light of evolutionary theory? Did Adam and Eve really exist at the origins of our history as a biological species? There are many others.

St. Thomas Aquinas (1225?–1274) was a Dominican friar, a priest, a philosopher, a theologian, and a Doctor of the Church. He is often called the "Angelic Doctor," not only because of his purity of heart but also because he wrote extensively on the angels. He is known best for his two summaries of theology called the *Summa Theologiae* and the *Summa contra Gentiles* and his commentaries on Aristotle's writings and on sacred Scripture.

It is universally acknowledged that Aquinas was one of the most brilliant and influential thinkers in the history of the

Catholic Church. It is not surprising therefore that Vatican II in its decree on priestly training proclaimed the following: "[I]n order that they may illumine the mysteries of salvation as completely as possible, the students should learn to penetrate them more deeply with the help of speculation, under the guidance of Aquinas, and to perceive their interconnections."[1]

As a team of Dominican friars and scholars committed to the preaching of the Gospel, we are convinced that the Thomistic intellectual tradition grounded in the philosophical and theological synthesis of Aquinas can still provide insightful and compelling responses to the disputed questions raised by evolutionary theory.

From our pastoral experience, we have discovered that Catholics and other Christians are frequently surprised by the novelty and brilliance of Thomistic responses to these disputed questions in evolutionary theory. Often, this is the case because these answers transcend and reconcile the dichotomies —for instance, the oft-cited dichotomy between chance and design—that shape the contemporary science and religion debate. God designs with chance! Unfortunately, the Thomistic responses to these disputed questions in science and religion are neither well known nor well understood.

To remedy this, we have written this book and inaugurated its companion website (www.thomisticevolution.org) with a series of responses to disputed questions that cover the topics we have encountered most frequently in our conversations with believers. Though the chapters can be read individually, we have intentionally put our answers in an order that systematically reveals the theological vision of our brother, the Angelic Doctor.

The book is divided into four parts. In the first part, we begin by describing our overarching vision of how reason and

---

1.   *Optatam Totius*, §16.

faith come together as two wings of the human spirit which rises to the contemplation of truth.[2] In the second part, we then move to several chapters that focus on disputed questions that examine the existence and nature of God without appealing to divine revelation. These philosophical questions are at the heart of the field of inquiry called natural theology. We then move to the third section of the book that examines the scriptural foundations for the Church's understanding of God's creative action in the world. What did God tell us about his role in how the diversity of life on this planet came about? Finally, we conclude the book with a series of disputed questions that deal specifically with both the scientific claims put forward by evolutionary biology and the important theological task of putting them into conversation with the teachings of the Catholic Church.

---

2. See Pope St. John Paul II, *Fides et Ratio*, §1.

✣ ACKNOWLEDGMENTS

This book and its companion website (thomisticevolution.org) would not have been possible without the help of our creative designer, Fr. Antoninus Niemec, O.P., and our webmaster, Fr. Innocent Smith, O.P. We also thank Antonio Andreu-Galvan and John "Jack" O'Reilly, for their work with the Spanish parish inserts available on the website. Finally, we acknowledge the talent of Daniel Mitsui who is responsible for the commissioned artwork for the website. This project was initially developed with a grant from the Evolution and Christian Faith (ECF) Program of the BioLogos Foundation. It is currently supported by a grant from the John Templeton Foundation.

❧ CHAPTER 1

Faith and Reason: The Two Wings of the Human Spirit (I)

Faith and reason are like two wings on which the human spirit rises to the contemplation of truth.[1]

With this image of the two wings, Pope St. John Paul II summarizes two thousand years of Christian reflection on the relationship between faith and reason. The image is surprising. For according to the predominant mentality of our times, one must choose between *either* being a person of faith *or* being a person of reason. One cannot be both. That there is a dichotomy between the two is almost a given within the contemporary academy and within our society at large, even among many Christians.

However, the good news that we will illustrate through and with these chapters on evolution and the Christian faith is that one and the same God who created us with reason also gifted us with faith. No one has to choose between the two. Everyone can have both. Faith and reason are meant to work together.

Why is the view that one must choose between either living a life of a faithful and devout believer or leading the life of an intelligent and enlightened adult, so widespread today? The answer is complex, but I will give five reasons here.

1

The first reason is that, at face value, faith and reason can appear to be opposed to each other. Because of the appearance of conflict, human beings had to learn gradually how these two ways of contemplating truth could work together. The integration between the two does not happen by nature but by nurture. For this reason, it took the Catholic Church well over a thousand years from her founding by the Lord Jesus to learn and then to show the Christian people and the world how faith and reason can come together harmoniously.

Furthermore, even after the first thousand years of development, one of the lessons learned was that the task of integrating faith and reason always continues in the Church. It is a task always before the Church due to new developments in mankind's ways of thinking and of understanding reality.

It is widely acknowledged that this task of synthesizing faith and reason was brought to unprecedented heights by the end of the thirteenth century, especially in the work of St. Thomas Aquinas. It was the accomplishment both of the Church Fathers, who lived in the first six hundred years of Christianity, and of the Scholastics, who lived between 1000 and 1300 A.D. However, this synthesis between faith and reason was fragile and vulnerable to fracturing. And it has been continually challenged and attacked ever since.

A second reason why faith and reason are widely perceived to be in conflict today is sin. The *Catechism of the Catholic Church* defines sin as follows:

> Sin is an offense against reason, truth, and right conscience; it is failure in genuine love for God and neighbor caused by a perverse attachment to certain goods. It

---

1. Pope St. John Paul II, *Fides et Ratio*, §1.

wounds the nature of man and injures human solidarity. It has been defined as "an utterance, a deed, or a desire contrary to the eternal law."[2]

Because of sin, human beings are prone to all forms of disintegration. Sin tears apart things that otherwise would go together peacefully in our lives. For example, the male and the female naturally form a complementary pair. In principle, a man and a woman are capable of living together in love, joy, and harmony. But sin has introduced many tendencies into the lives of men and women that incline them to treat each other in destructive and harmful ways. Sin is the origin of the mutual suspicion, the lies, and the discord that we often see in relationships.

Similarly with faith and reason. Although God meant both these ways of contemplating truth to work together, there are many sinful tendencies that make the work of integrating faith and reason difficult. The tendency to reject God's truth when it is difficult to live or understand; the tendency to take the world into our hands and dominate it for our own plans and purposes; the tendency to refuse to depend upon God from our hearts for a truth which is genuinely beyond our powers of direct verification; and the tendency to despair over the difficulties of working through all the many perplexities about God and the world. And then there is the tendency to despair over knowing truth itself. In a fallen and a sinful world, these are only some of the tendencies that tempt us to give up on the arduous task of reconciling faith and reason.

A third reason is historical. The sixteenth century was a time that placed an unusual number of perplexing social and intellectual challenges before the Church. The invention of the printing

---

2. *Catechism of the Catholic Church* (*CCC*), §1849.

press, the Protestant reformation, the discovery of "the New World," the rise of modern science, the growing awareness of the great diversity of world religions, and the development of new philosophies. These are only some of the historical developments that tested and continue to test the synthesis of faith and reason accomplished by the Fathers and the Scholastics by the end of the thirteenth century.

Moreover, as the Church herself was learning to deal with these new intellectual challenges, various elements of European intellectual life gradually drifted away from her orbit. As Europe became increasingly secularized, it became increasingly common for people to think that one must choose between belief and thought, between religion and science, between a rational life and a spiritual life, because they did not accept the guidance of the Church in seeking a harmonious whole. These dichotomies come down to us today as cultural givens, a kind of counter-tradition to the tradition of the Church.

A fourth reason is found in our contemporary culture. Our contemporary culture has an extremely impoverished understanding of what faith is and of what reason is. On the one hand, faith is equated with religion. It is commonly thought to be nothing but feelings about certain matters. It is a set of feelings about life, meaning, values, and God.

At other times, faith is understood to be a set of private convictions about these matters, but not convictions based on *evidence*. Rather, to many contemporary minds, faith has little or nothing to do with *truth*. A person's faith cannot be said to true or false. At most, a faith conviction is true *for this individual*, i.e., it is his belief. But the belief itself cannot be simply *true*.

Reason, on the other hand, is equated with science. It is understood to be thinking based on experiment, critical analysis, and evidence. The results of science, it is widely thought, are verified facts and publicly accessible truths. In fact, to many contemporary minds, only the results of science are really verified facts or publicly accessible truths. For contemporary people who think that faith is merely a matter of feelings that have nothing to do with truth, it seems *obvious* that faith and reason either have nothing to do with each other or have to be in conflict.

The final reason involves certain movements in contemporary society. In a society where faith and reason are not integrated and where it seems that one must choose between the two, many people choose one or the other with a conscious exclusion or rejection of the others. Some people choose to fly by reason alone, and they reject faith altogether. Most commonly, they claim to live on the basis of science alone. This choice is often called *rationalism* or *scientism.* Its motto could be, "Forget faith. Reason alone is the guide to life."

Other people choose to fly by faith alone, and they reject reason in some serious way. They may sincerely and deeply believe the Bible or some other religious text, but as is often the case, they refuse to ask hard questions about the meaning and interpretation of this sacred text. Thus, they often refuse to accept well-established results of modern science because they think that these scientific claims would undermine their faith. This alternative is often called *religious fundamentalism.* Its motto could be, "Don't think. Just believe."

These contemporary tendencies take the *perceived* conflict between faith and reason and harden it into actual opposition between fixed positions. The widespread presence of rationalism/scientism and fundamentalism only compounds the difficulties of learning to reconcile faith and reason in a life lived harmoniously.

In sum, it takes time, teaching, and effort to learn how to integrate faith and reason. Our sins and our weaknesses make it difficult to learn to fly with both wings of the human spirit. The history of the last several centuries has moved our civilization away from an intellectual synthesis that shows us how this can be done. Our contemporary culture does not teach people how to fly with both wings, and it is populated by vocal minorities who confuse people about even the possibility of synthesizing faith and reason. Is it any surprise that for nearly all people today it *seems* that faith and reason are opposed to one another?

The purpose of these chapters on evolution and Christian faith is to show men and women of good will how to grow out of the dichotomies between faith and reason that have crept into our society. In it, we hope to show how the Catholic Church's intellectual tradition, and especially the synthesis of St. Thomas Aquinas, offers our contemporary society a powerful and potentially life-shaping wisdom that is ever ancient and ever new.

Our plan is to investigate the topics raised by creation and by evolution in order to demonstrate how the Catholic Church's understanding of faith, reason, scripture, man, grace, the fall, and redemption through Jesus Christ can cohere marvelously and profoundly with a contemporary scientific account of the origin of life, the development of species, and the origins of our species.

We want to illustrate our claim that one does not have to choose between two stories of the world, the world according to the Bible and the world according to science. Rather, there is one world. It is God's world. And it is a good world. Created in love and wisdom, fallen and redeemed, God and the world are knowable by faith and reason together.

James Brent, O.P.

❧ Chapter 2

What Is Reason?

In the previous chapter, I presented faith and reason as two wings on which the human person is meant to soar unto the contemplation of truth. Human beings can learn to fly with both wings, but in our contemporary culture, people tend to think they must choose between faith and reason. One of the main reasons for this is that our contemporary culture operates on a superficial notion of what reason is. In the popular mind, *reason* tends to be reduced to *science*, i.e., modern, experimental science. To identify reason with science is a position commonly called *scientism*. The first step in learning to fly with both wings is learning to see through scientism to a deeper understanding of reason.

Let us begin with a definition of science. The Science Council of the United Kingdom proposes the following definition of science on its website: "Science is the pursuit and application of knowledge and understanding of the natural and social world following a systematic methodology based on evidence." It is likely that scientists from throughout the world would agree with this definition.

The website goes on to say that this systematic methodology includes objective observation, measurement and data, evidence, experiment as a benchmark for testing hypotheses, induction, repetition, critical analysis, and verification. The primary examples of scientific research are clear enough for us to have an idea of scientific methodology.

Because contemporary people tend to reduce reason to science as so defined, truth too, tends to be reduced. What is truth? In our society, when most people hear the word "truth" they automatically understand it to mean: scientifically verified facts or information. A deeper understanding of reason, we shall see, leads to a deeper understanding of truth.

There is a major price to pay for holding a scientistic account of reason and truth. For if reason is limited to science, and truth is limited to scientifically verified facts, then it very much seems there are no true or rationally verifiable answers to questions about God, about morality, or about the meaning of life.

First, concerning God. According to the definition of science given by the Science Council, science deals with the natural or social world. God, however, is beyond the natural or social world. Science, therefore, by its very definition cannot settle the question of God's existence or attributes. If reason is limited to science, then by definition, reason also cannot settle these questions.

Second, concerning moral claims. Given the methodology of science outlined by the Science Council, moral questions cannot be settled by science. This is more controversial, since there are some scientists and philosophers who want to claim that science can make moral claims and settle them. But the more predominant position is that science and morality are completely separate. The former deals with facts, using its own proper methodology, while the latter deals with values and has its own proper methodology.

Influenced by the claims of the 18th century British philosopher, David Hume, many ethicists hold that one can never move from factual claims to value claims. One can never go from a claim about the way the world is, to a claim about the way it ought to be. Furthermore, many philosophers and scientists hold that the method of science is powerful and makes such rapid progress precisely because it steers clear of value claims or claims about the way things ought to be. For these reasons, many people (and certainly the popular mind) are persuaded that moral precepts or moral codes are outside the range of scientific verification. Again, if reason is science, and science cannot settle moral questions, then reason cannot settle moral questions.

Third, questions about the *meaning* of life or *meaning* of reality cannot be settled by science. For science, it seems clear, is not designed to raise such questions, propose answers to them, or verify those answers. If we ask a physicist, biologist, or chemist to describe life, we may get an answer. But it is not the physicist speaking *as a physicist* or a biologist *as a biologist* who proposes that answer. Psychologists sometimes raise the question of meaning and argue that the issue is of central importance to human health and therapy, but even the psychologists who explore the issue tend to claim that meaning is utterly subjective. In other words, there is no objectively true and scientifically verifiable answer to the question of life's meaning. Rather, each person constructs his or her own meaning like a spider spinning a web out of itself. If science is all we have to go on, then what shall we say about the meaning of life? There is no scientific answer to the question, and if reason is limited to science, there can be no rational answer.

On the scientistic account, reason knows nothing about God, can settle nothing about morals, and knows nothing of life's meaning. Unfortunately, these are not merely theoretical implications. Large numbers of people in our society, especially many students on our college campuses, go through life with the assumption that there is no rational and objective answer to the questions of whether or not God exists or of what his attributes would be. It is often taken for granted that there is no way to settle moral questions other than with a vote. Many are convinced that there is no universal, true, rational, accessible, verifiable answer to the question of what life is all about. The reduction of reason to science has left multitudes of people without a moral or spiritual compass. Their lives become subjective journeys where they alone determine where they are going. When this Godless, guideless, meaningless form of existence is put forward as a philosophy, the philosophy is called *nihilism*. It is a belief in nothing as ultimate. In his encyclical *Fides et Ratio*, Pope St. John Paul II described nihilism as a significant challenge for our times. He realized that a common and shared civilization could not be sustained on such a basis.

Many secular and non-Christian philosophers have come to the same view. Years before Pope St. John Paul II wrote his encyclical, one of the most prominent philosophers of the twentieth century, Edmund Husserl, foresaw a major crisis coming for western civilization as it tried to run on the basis of modern science alone.

However, there is no need to embrace such a narrow or reductionist account of reason. Science is a good way to know facts, but *science is only one way of knowing among many.* There are other ways of knowing. Consider, for example, literature. Reading Shakespeare offers an abundance of insights into profound truths about life. "What's done cannot be undone."[1] When we read that sentence in light of the text and our experience of life, it encapsulates many deep and objective truths that science could never verify.

---

1. *Macbeth*, Act 5, Scene 1.

There is another account of reason. It is more ancient, richer, and more open to reality as a whole. Reason, on this account, is *sapiential*. Reason here is the capacity for *wisdom*. Wisdom is an all-embracing understanding of reality as a whole in light of ultimate causes, especially in light of the *end* or *goal* of all things. In order to be capable of such wisdom or such an all-embracing understanding of reality, reason must be receptive to reality in *all* of it aspects: the quantifiable and the non-quantifiable, the measurable and the immeasurable, the observable and the non-observable, the tangible and the intangible, and the sensible and the intelligible.

Reason is not prejudiced against God, against moral truth, or against the meaning of life. Rather, reason understood as wisdom is open to such questions, seeks answers to them, and is capable of finding true answers to them. Wisdom does not deny modern science but goes beyond it. Modern science alone, especially when it denies that there is an end or goal to all things, can never deliver wisdom. Something more than science is needed for wisdom, and something more than science is available. Reason is this something more, and reason is capable of an open and fruitful inquiry into reality as a whole. This entire book on Christian faith and evolution is a task that illustrates the work of reason understood as wisdom. It seeks to bring a synthetic whole to one's vision of the world by seeking the causes for all that we can experience and understand.

The Church understands reason in this ancient, richer, and more open sense. Because the Church understands reason as wisdom, the Church teaches that human reason is capable of arriving at some limited knowledge of God's existence and attributes, at solid insight about moral truth, and at a deep grasp of the meaning of life. Furthermore, reason understood as wisdom is open to receiving a divine revelation if God should so deign to give us one. It is even capable of detecting signs of authentic divine revelation and distinguishing this authentic revelation from counterfeit rivals.

Nihilism is not a necessary attribute of human existence! There is objective truth that we can attain as we seek the answers to the most important matters in life. However, let us be aware that our aptitude to know these things is limited in many ways. Seeking truth is often difficult and one must persevere if one wishes to attain it. But the truth is there. It is real.

On the sapiential account of reason, truth is more than just scientifically verified facts. We can distinguish between Truth with a capital "T" and truth with a small "t." Truth with a capital "T" is the *meaning of reality as a whole*. Truth with a small "t" is one of the truths or facts about the world. Science helps to verify many facts about the world, it delivers many truths, and no wise person wants to deny solidly established scientific facts. But if reason is something more than just science, if reason inherently seeks Truth and wisdom, then reason is inherently driven to an understanding of it all—the meaning of reality. And if reason questions the meaning of all things, perhaps God in his goodness may speak to our questions with answers from on high. What then, would faith be, except believing God's answers to the questions that we ask with the reason that he has given us?

James Brent, O.P.

❧ Chapter 3

What Is Faith?

Faith and reason are like two wings on which the human person soars unto the contemplation of the truth. Such is the good news of this book. In our last chapter, we addressed the question of what reason is, and in this chapter, we address the question of what faith is.

In our last chapter, we said that reason is something more than science. Reason is the capacity for wisdom, and wisdom is an all-embracing account of reality as a whole in light of the highest end or purpose of things. On this account, in order for human beings to be fully rational, we must be receptive to reality in *all* of it aspects: the quantifiable and the non-quantifiable, the measurable and the immeasurable, the observable and the non-observable, the tangible and the intangible, and the sensible and the intelligible. The Church's account of faith begins from this understanding of reason as sapiential. When human reason is open to reality in all of its fullness, then reason questions and searches for Truth with a capital "T," i.e., the *meaning* of reality as a whole. When God comes to meet our questioning and searching for Truth, and address his own answers to us, then human beings are confronted with divine revelation and called to faith.

In contemporary popular contexts, faith is understood in a bewildering variety of ways. Sometimes, the word faith is used to talk about a person's perspective on ultimate questions. At other times the word faith means a bunch of feelings or convictions that one has about things, i.e., one's deepest values. At other times the word faith is practically synonymous with one's "philosophy of life." Or faith can be used to talk about any cause that one really believes in and advocates. Because faith is used in such a vague and general way, it is common to hear people using expressions such as "the Buddhist faith" even though Buddhism is not at all a faith in the Christian sense of the term, and Buddhists (rightly) want to distinguish their religious views from being a faith in the Christian sense.

It was St. Augustine who gave to the Catholic Church an account of what faith is that has become standard up until our own day (his view is presupposed by Vatican I and II, the *Catechism of the Catholic Church*, and St. John Paul II's encyclical on faith and reason, *Fides et Ratio*). St. Augustine, following both the New Testament as well as standard word meanings in ancient Greek and Latin, understood faith as *believing something on the word of a witness*. It is believing something based on the authority of another. The New Testament is full of talk of testimony, of testifying, and of bearing witness to the truth of Christ's life, death, resurrection, grace, and presence. Faith is welcoming this testimony, accepting it, and believing it.

St. Augustine realized that when understood as believing something on the word of another's witness, faith in general is inherently reasonable. All human beings naturally live by faith in other human beings. It is quite impossible for anyone to go through life without any faith at all. For we all take many things on the word of other people, and we cannot but do so. Augustine realized that nearly all of his historical beliefs, including his belief about where he was born and who his father is, were matters of faith. Contemporary epistemologists confirm Augustine's point, and point out how the majority of our daily beliefs are matters of faith in the sense of trusting someone else's say-so. How many people can really *prove* that water is composed of hydrogen and oxygen? How many people can really *prove* that George Washington was the first president of the United States? Most people take it on faith based upon their parent's or teacher's say-so.

Even though faith in other human beings is natural and reasonable, Christian faith is something more than just natural, human faith. The same God who created human beings as rational beings in search of the meaning of reality as a whole, comes to meet the questioning of man by bearing witness to himself and his plan for the world. God explains Himself to us. How? Through the words of the prophets, the words of Jesus of Nazareth, the words of the Apostles, and the words of the authors of the Scripture. All of these human words contain and communicate the Word of God. Or so Christians believe. Christian faith is a special gift of God to believe all of this testimony "not as the word of men, but for what it really is, the word of God."[1] In brief, faith is welcoming the Word.

How does one receive the gift of faith? The answer has three parts.

First, "faith comes from hearing."[2] Ever since the days of Jesus, the apostles, and the evangelists, the Church has gradually welcomed, received, and believed their words as the Word of God. The Church has assimilated the person and the message of Jesus and in turn preaches and proclaims it: *Jesus is Lord!* The Church goes through history vouching for the truth of this word to all who will listen. What we hear are the words of human

---

1. 1 Thess. 2:13.
2. Rom. 10:17.

beings in the Church setting forth the person and message of Jesus as the Word of God.

Second, as the Church goes about preaching, "God also bore witness by signs and wonders and various miracles and by the gifts of the Holy Spirit distributed according to his will."[3] God knows of our fear of deception. He knows of our felt need for confirmatory signs that what we are hearing from the Church is not simply the words or ideas of human beings but is in fact the Word of God. The preaching of the Church is surrounded by and shot through with an abundance of signs and characteristics that distinguish it from all merely natural phenomena. Miracles, healings, and transformed lives are but one type of sign. The Church herself is a sign. The signs, in fact, are so many, so varied, and so numerous that an entire academic discipline is devoted to studying and setting forth the signs, namely, fundamental theology.

Third, "no one can say that Jesus is Lord except by the Holy Spirit."[4] Even though a host of signs confirm the Church's proclamation as the Word of God, people are not called to believe-because-of-signs. Rather, we are called to a free, simple, childlike faith in Jesus as risen Lord and Savior. It is not an inference from signs that moves us to believe what we hear, but an inward touch of the Holy Spirit. The Holy Spirit inclines our hearts to trust God for the truth of what is set forth by the Church. Thanks to the inspiration of the Spirit, we give a simple assent to it all—a simple yes to the whole testimony of the Church without fear of falsehood or error. Signs are there for critical reflection upon our faith, not for the production of it.

The words we hear, the signs we see, and the Spirit's touch within us, together show each one of us how right it is to believe.

---

3.   Heb. 2:4.
4.   1 Cor. 12:3.

St. Augustine, confronting the words of the book of Genesis wondered whether what he was reading was really the Truth. Here is how Augustine describes the touch of the Spirit within him. In his *Confessions*, he writes:

> May I hear and understand how in the beginning you made heaven and earth (Gen. 1:1). Moses wrote this. He wrote this and went his way, passing out of this world from you to you. He is not now before me, but if he were, I would clasp him and through you beg him to explain to me the creation. I would concentrate my bodily ears to hear the sounds breaking forth from his mouth. If he spoke Hebrew, he would in vain make an impact on my sense of hearing, for the sounds would not touch my mind at all. If he spoke Latin, I would know what he meant. Yet how would I know whether or not he was telling me the truth? If I did know this, I could not be sure of it from him. Within me, within the lodging of my thinking, there would speak a truth which is neither Hebrew nor Greek nor Latin nor any barbarian tongue and which uses neither mouth nor tongue as instruments and utters no audible syllables. It would say: "What he is saying is true." And I being forthwith assured would say with confidence to the man possessed by you: "What you say is true."[5]

Augustine heard the words of Scripture. The Spirit touching his heart bore witness to the Word of God contained therein. Such is the gift of faith. The gift of faith carries with it the "proof of things not seen."[6]

---

5. *Confessions*, Book 11, III (5).
6. Heb. 11:1.

When the human being has received by faith God's testimony of Himself and his plan, the drive to understand the meaning of reality as a whole can come to a new level of fulfillment. For the testimony of God provides answers to the most profound human questions. The question of God's existence and attributes, the question of God's providence, the question of evil and suffering and death, the question of the afterlife, the question of personal goodness and worth and the good life, the question of the limits of reason, the question of love and meaning, and many more questions all receive answers in the testimony of God received by faith. Those answers are open to reflection and call for thinking. Faith is pondering those answers with assent, thinking them through; and, when such pondering and thinking is carried out systematically, the result is theology.

A good and solid theology does not deny the truth of the human sciences but takes into account all well-established truths from any discipline whatsoever. The Word of God illuminates all such truths in a higher light, in the light of God and his plan for the world. Such is the intention of this book. Our hope is to show that the science of biology and the testimony of God together form a coherent and profound answer to the question of the human being, our life in this world, and the meaning of it all.

James Brent, O.P.

CHAPTER 4

Faith and Reason: The Two Wings of the Human Spirit (II)

When faith and reason are both properly understood, it is much easier to learn how to fly with both wings of faith and of reason together. Birds, we observe, *learn* to fly gradually and by practice. So, too, human beings *learn* how to fly with faith and reason gradually and by practice.

In order to facilitate that learning, we now present seven ways in which faith and reason relate to each other. Each of these seven ways also reveals an ongoing task for the Church in order for us together to cultivate an intelligent faith and faithful intelligence. In each of the following numbered points we name the relation, define it in italics, clarify it with further comments and examples, and discuss the ongoing task it sets before the whole Church.

1. Consistency. *Right faith and right reason are logically consistent with each other.* Whatever God has *in fact* revealed for acceptance by faith and whatever has been *genuinely* demonstrated by reason do not and cannot contradict. For if God reveals something to be accepted by faith, then it is true. And if something has been genuinely demonstrated by reason, then it too is true. Since no truth can contradict another truth, what is held by a right faith and what is demonstrated by right reason cannot contradict one another.

The difficulty is that faith and reason sometimes *seem* to contradict. In such cases, we know that the appearance is due either to a faulty use of reason or to a misinterpretation of divine revelation. It would be a faulty use of reason to say that the universe is in a steady state without beginning or end. It would be a misinterpretation of divine revelation to say that God has revealed that the sun is the center of the universe. Because there can be faulty uses of reason or misinterpretations of divine revelation, it is important to say that *right* reason and *right* faith are consistent.

At any one point in time, however, the Church faces a host of apparent contradictions between faith and reason. We can call these apparent contradictions simply *difficulties*. These difficulties are valuable precisely because they tell us there is either a faulty use of reason or a misinterpretation of divine revelation somewhere in our thinking, and the difficulties call us to think more deeply, do more research, draw distinctions, and grow in our understanding. Aristotle thought that the collecting of such difficulties, and the sorting out of such difficulties, was at the heart of the *process of rational discovery* and *growth in understanding*. Contemporary philosophers of science have shown how a great deal of scientific research consists mainly in accounting for aberrations and anomalies in current research paradigms. So it is too for Christians. Understanding grows by gradually resolving difficulties between faith and reason.

There are many contemporary difficulties. Evolutionary biology seems to affirm that man is not a special creation by God but just the product of a meaningless chance process, while sacred Scripture seems to affirm that man is a special creation and not the product of a meaningless chance process (or not *just* the product of meaningless chance). It has been claimed that historical critical exegesis of Scripture seems to affirm that Jesus of Naza-reth did not claim to be God, although divine revelation received in faith according to the understanding of the Church seems to affirm that Jesus of Nazareth did claim to be God. According to divine revelation traditionally understood, the human person has a soul that survives death. According to the vast majority of cognitive science researchers and philosophers of mind today, the human person does not have a soul at all. In all of these cases, lay-ers of lesser *difficulties* need to be resolved. Naming the apparent contradictions, elaborating the issues they involve, and seeking to resolve them are a vast task at hand.

2. Support. *Right reason can demonstrate many truths about God and support some of the things we believe by faith in God's revelation.* The Church claims that human reason, rightly used, is capable of giving solid arguments for the existence of God as well as various divine attributes. In so doing, human reason can also account for how its own language applies to God. Human reason can also provide good arguments for the immateriality and immortality of the human soul. Furthermore, human rea-son can also provide many arguments that corroborate or con-firm that what we believe is a divine revelation is in fact a divine revelation (as opposed to a lie, fraud, myth, or psychological dis-order). In all these ways, right reason supports what we believe by faith.

The Church also understands that not all human beings are equally capable of doing all the intellectual learning needed to elaborate or fathom all such arguments. The task, rather, falls to a few who are gifted with the aptitude, time, and leisure for such studies. Nonetheless, there are in the Church some people with a calling to take up just this task of searching into reasons in support of the faith, and the Church exhorts all to develop their own personal intellectual potential to the full.

3. Defense. *Right reason can refute objections brought against things we believe by faith*. Skeptics are often not content with requesting evidence that supports Christian beliefs, but they often advance arguments attempting to show that Christian beliefs are false. Human reason, sometimes proceeding as philosophical reason and sometimes proceeding in the manner of other disciplines such as history, archaeology, biology, and philology, is able to show that such arguments fail to arrive at their conclusions.

For example, it is commonly said that the Incarnation of God in the man Jesus of Nazareth is impossible, or the doctrine of the Trinity is incoherent, or that the resurrection of Jesus from the dead could not have happened, or that transubstantiation (the real presence of Christ in the Eucharist) is absurd, or that the doctrine of papal infallibility is provably false on historical grounds. All of these claims are advanced with arguments, and sometimes apparently compelling arguments.

Human reason does not always *easily* find the fallacy in such arguments. The current state of the evidence may provide what seems to be deep support for positions incompatible with a particular Christian belief. But nonetheless it falls to reason to answer these objections if only by undertaking new research with a view to refuting them in the future. The task at hand is to gather these objections, humbly give them a fair hearing, humbly welcome what truth they hold, and answer them without a proud and polemical spirit.

4. Illumination from below. *Right reason can discover many truths that enhance our understanding of what we believe by faith in God's revelation*. Any truth knowable by philosophical reason or by human learning generally has the potential, either directly or indirectly, to enhance the believer's understanding of the mysteries of faith. There are many examples.

The classical principle that "all goods are either means or ends or both" is very illuminating for the faith. St. Thomas Aquinas understands our faith such that union with the Holy Trinity is the *end* of human life, and the Incarnation is the means to that union.

Aristotle's understanding of human nature as a body-soul composite possessing certain powers is a truth knowable by philosophical reason that helps us to say more precisely what we mean when we say that God became man in Jesus of Nazareth. Jesus had a human body and soul and all of the powers that go with the composite.

The distinction between substance and accident illuminates the mystery of the Eucharist. In the Eucharistic celebration, the bread and wine change in their very substance into the Body and Blood of Jesus, but the accidents of bread and wine remain.

The distinction between agent and instrumental cause illuminates God's relation to creatures. God is the primary cause of creatures who gives them being and influences their activities. Creatures are instruments through which God produces his effects. God uses rain to hydrate living things.

Historical and archaeological studies, too, have uncovered truths that illuminate our faith. To give a particular example, sound scholarship has established that "the Judaism of the [Christ's] day was familiar both with more generally formulaic confessions of sin and with a highly personalized confessional practice in which an enumeration of individual sinful deeds was expected."[1] This historical fact sets the context for the baptism of Jesus, and helps us to see how by his baptism Jesus chose to be counted among sinners as an anticipation of his cross and also in fulfillment of prophecy.[2]

---

1. Pope Benedict XVI, *Jesus of Nazareth*, trans. Adrian Walker (New York: Doubleday, 2007), p. 15.
2. See Is. 53:12.

Whether philosophical reason or historical reason or scientific reason, all truth gathered "from below" potentially serves to understand the mysteries of faith. It is the task of systematic theology to take from the truths discovered by right reason and use them to illuminate the riches of the faith.

5. Correction. *Right faith corrects many errors that reason commonly makes.* It is commonly acknowledged that the history of human learning is a history of errors and mistakes (though it is not just errors and mistakes). Human inquiry, therefore, stands to benefit from having a higher light or criterion to show such errors and mistakes for what they are. Now, as pointed out above, any purported deliverance of reason that contradicts what God has revealed cannot itself be true or a deliverance of right reason. For this reason, faith and revelation point out many errors of reason.

Standing on faith in the divine revelation given in Christ Jesus, as understood by the Church, one can know quickly that moral relativism is not true, materialist reductionist accounts of the human person are not true, that nihilism is not true. If a believer hears it claimed that the world is eternal, or that there is no afterlife for human beings, or human beings have no free will, or that human beings are *nothing but* the product of a meaningless, blind, chance process, then the believer's faith alerts him or her to the presence of error. Not all errors are detected so easily. Both a deep catechesis and deep personal holiness make believers more sensitive to more subtle errors.

Given the corrective power of the faith, the task at hand is to point out errors, especially the subtler ones, and to show what is compatible and incompatible with what God has revealed.

6. Illumination from Above. *Right faith allows the person to see all created things in relation to God.* If one believes that God exists, has created all things, and rules all things by his providence, then a wonderful form of contemplation opens before the believer. All things can be "read" in light of God.

For example, let us consider the distance of the earth from the sun. If the earth had been a bit closer to the sun, then the earth would be too hot for life. If the earth had been a little farther from the sun, the earth would have been too cold for life. How then shall we understand the fact, and the *meaning* of the fact, that the earth is approximately ninety-three million miles from the sun?

If we set aside the existence and providence of God, then the distance of the earth from the sun is just one more fact among many to be explained. But if God exists, created all things, and orders all things well by his providence, then the distance of the earth from the sun has a deeper significance or meaning—a sort of depth dimension. For the distance of the earth from the sun is now a fact *related to God*. Given that relation, it makes sense to say that the distance of the earth from the sun is just what God arranged it to be for human life to flourish on earth.

The existence and providence of God sheds light on the meaning of many more facts too, and perhaps on the meaning of *all* the scientific facts. The elements and their properties, initial conditions of the cosmos, the earth's initial chemical composition, the various cycles of water, nitrogen, etc., all of it has a depth dimension if God exists, created it, and orders it by his providence.

The Fathers of the Church were accustomed to "reading" the world this way—in light of God. So were Christian thinkers after them for centuries. One of the destructive effects of modern research and educational methods, which intentionally aim to be neutral about whether God exists and exercises providence over things, is that academically qualified researchers no long read the world in light of God or teach people how to read the world in light of God. This, we venture to say, is one of the contributing factors to the widely felt sense of the absence of God from our lifeworld.

The task at hand for us today is to relearn this form of contemplation that reads the world in light of God's existence and providence. Even more radically, the task is to read all things in light of the revealed mysteries of the Trinity, the Incarnation, and the Paschal Mystery.

7. Fulfillment. *Right faith provides answers to some of the most common, yet most difficult, questions raised by reason.* Human reason has many questions, but the history of thought proves that human reason faces great difficulties in finding shared answers to its own most profound questions. Philosophical questions about whether life has a meaning and what it is, whether there is a God, why God would permit evil, whether human beings have a purpose and happiness beyond life in this world and what it is, whether it is possible for universal justice ever to be secured in human society, and other such fundamental questions have been raised seemingly in all or most human societies. But answers to those questions seem also to elude human reasoning both with philosophers and with common people. The diversity of answers given, the mutual contradictions, the endless debates, and the shifting sands of time and culture make things difficult enough, but in our day, when positivism, scientism, relativism, nihilism, and anti-metaphysical postmodernism are influential, it is all the more difficult for people to avoid despair over coming to widely shared answers to a variety of significant human questions.

Faith in divine revelation, however, speaks to all of these questions and more. When one believes by faith the testimony offered by God, one receives answers to the questions of life. And those answers are susceptible to support, defense, and illumination by reason. The same answers serve also to correct and illuminate our own human learning.

The task today falls to Christian philosophers, theologians, and scientists to show precisely how divine revelation speaks to the deepest human questions and offers answers that form a compelling account of reality as a whole.

James Brent, O.P.

Chapter 5

Causality: What Are the Four Causes of Things?

One of the first things we learn about something is its name. We see this principle in toddlers who spend much of their time grappling with the names of things, whether they are dogs and cats, trucks and ponies, or Mom and Dad.

At some point, the simple task of identification is not enough to satisfy the child and parents start to face the dreaded question, "why?" This shift mirrors a shift in our own engagement with the world as we seek to understand it more deeply by turning from the mere fact of things to an understanding of their causes. Of course, the question "why?" can be asked and answered in a myriad of ways and it is worth taking a moment to probe its depth.

Take, for example, 8-year-old Junior approaching his loving Daddy, source of all knowledge and wisdom in his little world, and asking, "Why is the sky blue?" Dad might try to get off easy and simply answer, "Because the sun is out." Although even a child knows that the sky is only blue during the day, Dad's statement expresses an important point. The sky is not always blue, and the reason it is blue in the day but not at night is because of the sun. By its agency, as an intense source of light, the sun causes an effect in the sky that lights it up and makes it appear blue. The sun is an efficient or agent cause of the sky's blue color.

Unfortunately, Junior is not going to let Dad off that easy and persists asking, "Why is the sky blue during the day?" Realizing he's not getting off so easy, Dad replies, "Because the sky absorbs a lot of the light from the sun, and some of that light gets showered down on us." This second reply touches on what is known as "final causality," focusing on the natural tendency of the air. Analogous to the way we can explain our actions by referring to the goal or end we intend, we can explain many natural processes as a tendency towards a state of stability and rest. As the light from the sun excites the air and is absorbed, the air releases this extra energy to return, momentarily, to its relaxed state and some of that energy reaches us as light, the brightness of the sky. Once the sun goes down the air is able to simply relax in its natural rest state.

Undeterred, Junior presses on with a simple, "Why?" Dad, seeing that things are getting serious, answers, "Because the sky is a big layer of air with lots of little parts, little particles of air. Any particular piece of sunlight might hit any particular piece of air and send it off in a random direction, some of it coming towards us."

By appealing to the structure of the sky, as a diffuse gas, and the properties that follow from that structure Dad has laid out an argument from form. Broadly speaking, a formal cause describes the shape or organization of something and all the various activities that follow from it. While the form of air may simply be the arrangement of molecules, as well as the arrangement of the parts of those molecules, in some things, like living organisms, the form describes not only the arrangement of parts, its molecules, cells, and organs, but also the ordering of those parts to the good of the whole organism, and the activity of the organism as a whole.

Rather enjoying the worried look on Dad's face, Junior again replies with a further query, "Why is the sky *blue?*" Not willing to give up, Dad presses on, "Because the molecules of the various gases that air is composed of are much smaller than the wavelength of visible light, the shorter wavelength blue light is scattered much more frequently."

Here we have an argument from what air is made of, its matter. While the organization and activity of something is ultimately determined by its formal cause, the matter, or material cause, limits the potential forms the thing can take on. The density and temperature of air can be changed creating various weather patterns, but no natural process can force a bunch of air molecules to take on the form of a hippopotamus, however delightful that would be to Junior.

It may seem that Dad has made a valiant effort at explaining why the sky is blue to his son, providing a robust explanation from a scientific perspective and even touching on all the various senses in which the question "why?" can be answered. Junior, of course, may not be satisfied and has had the upper hand in this exchange, being limited only by his stamina and his attention span.

Note that Dad still has a wealth of causal explanations available as he can dive deeper into each of the four types of causes he touched on briefly, efficient, final, formal and material. He could turn back to efficient causality in order to trace the chain of causes that explain where the photons emitted by the sun come from or the various sources of the air in the atmosphere.

Alternatively, Dad could peel back the layers of material and of formal causality in the air. He could describe the structure and distribution of the various molecules that make up air, even delving into what the individual molecules are made of, digging down to the most fundamental structure of all matter. In the realm of final causality, Dad could place the basic scattering process of air in the atmosphere into the wider picture of order in nature, and its role in protecting the surface of the earth from more harmful radiation, regulating the air temperature, and producing the weather and climate that allow for higher levels of order and structure that we see in the ecosystem.

As we leave Junior and Dad to probe the depth of scientific understanding of the beautiful blue sky, and perhaps even some deeper philosophical truth, it is worth making a few more general comments on the idea of causality.

First the idea of the four causes, formal, material, efficient and final, dates back to the ancient Greek philosopher Aristotle (384 B.C.–322 B.C.) and is the broad structure in which St. Thomas Aquinas approaches explanations in nature. For them a cause is an explanation for how a thing comes into being, how it remains in being, and eventually, how it ceases to be, by becoming something else. This classical understanding of causality, where causes explain the being of things, is not the structure with which modern science generally approaches its work, but it is not contrary to modern finding and methods and any fully satisfying scientific explanation will touch on all four classical causes.

Much has been written and discussed on the role of these four types of causes over the past twenty-three centuries since they were first posited by Aristotle, and our somewhat simple example of the blue sky is far from a complete or satisfactory treatment of the topic. Nevertheless, it reveals the great value in thinking about the world in these causal categories, in particular when they are seen in the proper depth that our example tries to draw out.

Two of these causes, efficient and final, deserve a closer look because of their importance for the disputed questions that follow.

First, in efficient causality it is often possible to point to one particular cause that is the immediate agent through which the effect is made real. However, we can (and often do!) ask about the cause of the agent who is itself a cause. Sometimes, this is a matter of going back in time through a chain of events, as we when trace the efficient causality of sunlight back to the sun or back to the nuclear fusion events at the sun's core.

Other times we can see that the immediate agent is directly moved and powered by a moving agent, as when a saw cuts wood because it is moved by my hand. In this second case we find an example of instrumental causality where something acts as a cause, but only because it is empowered to act as such by another higher cause. In our example, the saw is an instrumental cause because it only acts, cuts wood, while being moved to do so by me.

Second, in our example of the blue sky, the final causality Dad describes is fairly simple: It involves the return of an excited molecule to its stable rest state. While this aspect of stability is the basis and root of final causality, we can see even more clearly a sense of order that arises from it in more complicated things. In living things in particular we see a tendency not towards any state, but towards states of perfection. An acorn tends to become a full-grown mature oak, a stable state capable of producing more acorns and oaks. A puppy tends to become a full-grown mature dog which in turn can produce more puppies and dogs.

We can see an even higher aspect of final causality in ourselves, when we consider the ordering principle of our own actions. We act with particular goals in mind, which are particular stable states in ourselves and in the world, which we think will make us happy. We look at the world and try to change it in little ways to be better for us. This aspect of final causality grounds the world of ethics and morality, which is rooted in the basic idea of tending toward a stable and perfective state.

Sometimes when people hear the phrase final causality, which is also commonly referred to as teleology, they assume that it refers to its highest form, the imposition of an external will on things. To many it seems like this is contrary to the very goal of our study of nature, the study of the inner working of the world around us. If teleology is only viewed as the external imposition of an intelligent will, they are right that it is contrary to this study.

However, at its root, teleology begins with the basic internal tendency of things to move towards particular stable states by their nature. Indeed, if it were not for this basic internal tendency of natural things to move towards stability, the whole project of science would be impossible because there would be no consistency or order to make nature intelligible.

Thomas Davenport, O.P.

❧ Chapter 6

The Existence of God According to Reason

In St. Paul's letter to the Romans, we read: "[E]ver since the creation of the world his invisible nature, namely, his eternal power and deity, has been clearly perceived in the things that have been made."[1] Here St. Paul sums up several passages of Old Testament wisdom literature.[2] The wisdom literature affirmed that the existence of God can be known not only by the divine revelation to the chosen people, but by human reason contemplating the world of nature all around us.

When this teaching was received by the Church, the early Christians observed that indeed there is a general consensus of the nations to the existence of some supreme Deity, and that some of the philosophers had offered arguments for the existence of God. As time passed, Christian theologian-philosophers then further developed those arguments for the existence of God and his attributes.

All of this has led the Catholic Church to teach definitively that "God, the first principle and last end of all things, can be known with certainty from the created world by the natural light of human reason."[3] In a world of much skepticism, this teaching can seem simply incredible. The purpose of this chapter is first to make several points about the nature and extent of the natural knowledge of God, then to offer a philosophical argument for the existence of God important for understanding the doctrine of creation and the meaning of evolution.

What, according to the Church, is the nature and extent of the natural knowledge of God?

The teaching on the natural knowledge of God is open to common misinterpretations. When the Church teaches that God can be known by the light of natural reason, she is not affirming that it is so obvious as to be undeniable to all people everywhere that God exists. She is not teaching that the existence of God can be verified by the methods of modern sciences. She is not affirming that there is one special philosophical proof or argument out there that will convince all people everywhere to know that God exists. What, therefore, is the Church saying?

The Church is presupposing our earlier account of reason. Human reason is open to reality as a whole in all of its aspects, seeks something beyond all the scientific facts, seeks to know the *meaning* of all things, and is capable of such knowledge to some extent. Reason is wisdom seeking. The Church also understands that, like all natural or human forms of knowledge, the knowledge of God *gradually develops* both in the lives of individuals and in societies, and consequently there are higher and lower forms and degrees of the natural knowledge of God depending on the conditions in which human beings live and develop. And the Church understands that there are many obstacles to the development of the highest forms of the natural knowledge of God. For these reasons, widespread disbelief in the existence of God is

1.   Rom. 1:20.
2.   Wis. 13:1–9, Ps. 19:1–4, Sir. 42:15–43:33.
3.   *CCC*, §36

consistent with the teaching that God can be known by the natural light of reason alone. For personal cognitive development, especially the natural knowledge of God, can falter in adverse conditions of life.

What are some of the conditions that favorably or adversely affect the natural knowledge of God? Where there is intellectual aptitude, interest, time for contemplation, a tradition of inquiring into the existence of God, and a will to worship God once known, such as in some of the ancient philosophical schools, the natural knowledge of God grows stronger and develops in its higher degrees and forms. Where these conditions are lacking, the natural knowledge of God—at least in its higher forms—flounders or is even opposed. Furthermore, the Church acknowledges that because of the fallen condition of the human race, the natural knowledge of God faces special obstacles in its development:

> Though human reason is, strictly speaking, truly capable by its own natural power and light of attaining to a true and certain knowledge of the one personal God, who watches over and controls the world by his providence, and of the natural law written in our hearts by the Creator; yet there are many obstacles which prevent reason from the effective and fruitful use of this inborn faculty. For the truths that concern the relations between God and man wholly transcend the visible order of things, and, if they are translated into human action and influence it, they call for self-surrender and abnegation. The human mind, in its turn, is hampered in the attaining of such truths, not only by the impact of the senses and the imagination, but also by disordered appetites which are the consequences of original sin. So it happens that men

in such matters easily persuade themselves that what they would not like to be true is false or at least doubtful.[4]

One could add that where skepticism about God, skepticism of metaphysics, or scientistic thinking dominate the atmosphere, where the practice of cultivating natural knowledge of God has been rejected, where the tradition of learning natural theology has been lost over the generations, or where a tradition of opposing it has been institutionalized and disseminated, widespread development of the higher forms of the natural knowledge of God is further impeded. For all of these reasons, one can say with St. Thomas Aquinas that if the human race were left to itself, without any special revelation from God, only a few people, after a long period of time, and still with an admixture of error, would develop the natural knowledge of God in its higher forms.[5] His statement illuminates our contemporary experience of the widespread cultural dominance of atheistic naturalism or physicalism.

In order to clarify further the nature and extent of the natural knowledge of God, let us distinguish between three degrees of cognitive development in the natural knowledge of God. One degree of development is pre-philosophical, a second is imperfect or rudimentary philosophical knowledge of God, and a third is perfect knowledge or rigorous philosophical demonstration of God's existence. Each of these forms of knowledge differs in its intellectual sophistication and the extent to which it is found among human beings. The teaching that God can be known by the natural light of reason may thus be understood in terms of three claims:

---

4. *CCC*, §37.
5. *Summa Contra Gentiles (SCG)* I, ch. 5.

1. *All human beings have a pre-theoretical knowledge of God.* This sort of knowledge comes to be in all cognizant human beings more or less spontaneously as we live in and think the world. It is general and confused knowledge—so general and confused, so primordial in our experience, that one is not even necessarily even reflectively aware of knowing God. One knows God without realizing it, for one knows him by a name other than God. Aquinas describes two ways of having this pre-theoretical knowledge of God.

In the first way, St. Thomas Aquinas writes:

> To know that God exists in a general and confused way is implanted in us by nature, inasmuch as God is man's beatitude. For man naturally desires happiness, and what is naturally desired by man must be naturally known to him. This, however, is not to know absolutely that God exists; just as to know that someone is approaching is not the same as to know that Peter is approaching, even though it is Peter who is approaching.[6]

All human beings are aware of goodness in general or have a notion of what goodness is—even if they cannot philosophically define goodness. It is the same with happiness. We all have a notion of it even if we cannot say what it is. Furthermore, all of us know that goodness and happiness are real. For we aim at them by nature, and we expect our aim to succeed. Just by having the notion of what goodness is or what happiness is, and by knowing that goodness and happiness are there to be had, one knows God. We could say that one knows God by the name of goodness or by the name of happiness rather than distinctly as

---

6. *Summa Theologiae* (*ST*) I, q. 2, a. 1, ad 1; *SCG* I, ch 11.

God. Aquinas compares this sort of knowledge to seeing some-one afar off coming over a hill without knowing it is Peter com-ing over the hill.

A second way of having pre-theoretical knowledge of God is from world order. St. Thomas Aquinas writes:

> For there is a common and confused knowledge of God which is found in practically all human beings; this is due…to the fact that…humans can immediately reach some sort of knowledge of God by natural reason. For, when human beings see that things in nature run accord-ing to a definite order, and that ordering does not occur without an orderer, they perceive in most cases that there is some orderer of the things that we see. But who or what kind of being, or whether there is but one orderer of nature, is not yet grasped immediately in this general consideration….[7]

When human beings gaze upon the beauty, order, harmony of the world as a whole, they commonly form the judgment that "there must be *something* behind it all." In this knowledge, they know God by the name of *something*, but not yet distinctly as God. What that something is remains an open question to the inquirer, but one knows at a minimum that *something* is there to look into further.

Generally speaking, the pre-theoretical knowledge is deeply compelling, often virtually indefeasible in one's mind, a power-ful starting point of inquiry, recurring food for thought, and, coupled with the innate desire to understand, it could drive one to elaborate philosophical arguments for the existence of God

---

7.   *SCG* III, ch. 38.

as a way of trying to put into words what one knows in a more primordial way.

2. *Many human beings have an imperfect philosophical knowledge of God.* Although we all start with a pre-theoretical or general and confused knowledge of God, human beings cannot be satisfied with it. For all human beings by nature desire to understand. And we are all fallen as well. Because we are driven by our nature to understand, many people unfold what they already know in a general and confused way into clearer and more distinct understanding. Because we are fallen, however, human beings can also refuse or oppose this process of cognitive development and effectively deny at a higher level what we know at a lower level. For those open to knowing the existence of God more perfectly than in the pre-theoretical way, the process of cognitive development advances according to our differing intellectual aptitudes, various degrees of free time for thinking, and differing degrees of concern for intellectual penetration of the subject matter. Many people take first faltering steps at trying to articulate their general and confused knowledge of God in more theoretical statements and arguments.

Hence, it is common to come across popular arguments for the existence of God. Someone may say: "Everything has a cause, but the causes can't go back forever, so there must be a God." Another may say: "Whatever is designed has a designer, and the world is designed, so there must be a Designer." These arguments represent first (or second or third) attempts to express some of the deepest intuitions of human reason about the ultimate meaning of all things. They are rudimentary philosophical arguments, and at whatever degree of sophistication they are developed they are open to easy refutation by someone with just a little more philosophical skill or to rejection by someone who is less astute at philosophy.

An objector may point out that if everything has a cause, then God too must have a cause, so the argument raises a problem for the proponent. Someone may point out that to say that the world is designed in fact begs a big question. *Is* the world in fact designed? Does not that claim presuppose the existence of God rather than prove it? In each of these cases, the objector may be genuinely more intelligent, or more thoughtful, or more philosophical, or more educated than the proponent of the argument for God. This may cause skeptical objectors to draw wild conclusions like atheists are more intelligent than believers or there are no good arguments for the existence of God, and all sides can easily forget that a good argument can appear bad to one who is not very intelligent. In fact, most people —both proponents and opponents—of the existence of God are living and thinking in this second degree of development of the natural knowledge of God. Philosophical arguments for the existence of God, and objections to them, fall along a spectrum of philosophical sophistication and rigor. Rigorous philosophical demonstrations are for the few: the intelligent, the educated, the dialectically skilled.

3. *A few people have a perfect philosophical knowledge of God.* Here we would propose St. Thomas Aquinas as an example of someone who arrived at a radical and penetrating understanding of reality, and we offer two of his arguments for the existence of God for consideration. We ourselves acknowledge that our formulation and understanding of these arguments is subject to our own further cognitive development as we ourselves grow and seek to understand reality as a whole with greater insight and clarity. Rigorous philosophical demonstrations of the existence of God are, after all, one of the highest accomplishments of human reason, and it is likely that we still have some learning to do.

The first argument is from world order. Aquinas often mentions that the order, harmony, and beauty of the world is the starting point for all rational ascent to the existence of God. In one place,[8] he offers an argument for the existence of God that we formulate and update in our own terms as follows:

1. In the world of Nature, we find things of different natural kinds.
2. The many things of different natural kinds each act in different and sometimes opposing ways.
3. Even though the many things of different natural kinds act in different and sometimes opposing ways, the world of Nature is coordinated, harmonious, orderly. (It is ecosystematic.)
4. Therefore, there must be something responsible for the coordination and harmony of the many different things in nature.

Furthermore, it seems to us that nothing other than God, including chance, the laws of nature, the four fundamental forces, or evolution, is a good candidate for that cause that best explains the harmony and order of the world. We also find various stories about how the human mind "constructs" the order of the world to be epistemologically problematic and frequently self-defeating.

The second argument is from contingent being. Properly speaking, the term "contingent being" here does not mean, a dependent being. It means a being that exists but does not have to exist. Aquinas is well known for finding within contingent beings a real distinction between *what* it is and *that* it is, i.e., between its essence and its existence. What a dog is, and that the dog is, are not the same, and this holds for anything less than

---

8. *SCG* I, ch. 13, §35.

an absolutely simple being whose very essence is to exist. This argument best begins with a meditation on contingent beings.

All around us we find many things that exist but do not have to exist: the sun, the moon, stars, plants, animals, human beings, even the Earth itself. According to the best cosmology, once upon a time, these things did not exist, and so we know they do not have to exist. Each one essentially can fail at being an existing thing.

When we ponder things that exist but do not have to, things that essentially can fail at being, the mind naturally asks why they exist at all. What gives them being or upholds them in being? With this question in mind, we formulate an argument from contingency as follows.[9]

We can pick out any particular being, for example, this individual dog here. It exists, but it does not have to exist. *What* it is does not cause, account for, or guarantee *that* it is. Its own nature as a dog does not guarantee that it exists or will continue to exist. At any one point in time throughout its journey in existing, the dog can fail to exist. So, its act of existing, *that* it is, must be received or supplied from without. It requires a cause of its very act of existing.

The parents of the dog are not the answer. The parents may be responsible for the dog *coming to be,* but they are not responsible for the *being* of the dog. For the dog can continue to be, even after its parents pass away. The cause of the dog's coming to be, and the cause of its being, must therefore be distinct.

The matter of the dog is not the answer. The matter of the dog is just as much in need of a cause as is the dog, since the matter of the dog is the dog, and the dog is contingent. The matter of the dog is as contingent as the dog is. Furthermore, there is

---

9. Inspired by *SCG* II, ch. 15, §5.

nothing about *what* matter is to guarantee *that* matter is. This holds both for particular bits of matter, even fundamental particles, as well as for matter in general. There must be something outside the dog altogether that makes the dog to be.

Let us now ask about the cause of the dog's being: Is that cause a contingent being? If so, then that cause too must have a cause, and we ask the same question. Is that too a contingent being?

We must arrive eventually at a non-contingent being that is a pure source of existing, something that does not receive its act of existence but only gives acts of existence to other things. For it is impossible for one thing to receive existence from another *without coming to a source or origin of existence.*

The argument is even more radical when we consider Aristotle's statement that cause and effect are simultaneous in act. Just as the mirror now reflects the light that now shines on it, so too, contingent beings now display the existence they are now receiving from the source.

We present the argument in the following form:

1. Some beings are contingent beings.
2. Every contingent being has a present cause of its very act of existing.
3. It is impossible to proceed to infinity in a series of things each of which is the present cause of the very act of existing of the next.
4. Therefore, there must be at least one being presently giving the act of existing to (at least some) to contingent beings presently existing which itself is not contingent.

One last point is worth making. A non-contingent being could only be a being whose essence is to be. We cannot conceive of any other way for a being to exist necessarily. The argument thus arrives at a Being itself that can give being to the contingent beings around us.

This argument poses as many questions as it answers, and we are open to objections in order to grow in our own understanding of the truth.

One last question deserves raising. Why claim that either of these two arguments arrives at God? Why claim that the cause of all the order in nature is God or that the necessary being that gives being to others is God?

One answer is that the entity arrived at by both of these arguments matches the dictionary definition of God.

Another answer is that the Bible itself makes the connection between the entity arrived at in these arguments and the God revealed in the Bible. In many places, the Bible affirms that the one responsible for the order of the world is God,[10] and in one place the Bible, at least traditionally understood, affirms that "I AM" is one of the names of God.[11] Traditionally, this has been understood to coincide with metaphysical arguments for the existence of God. The philosophers came to the God who also came to Moses.

James Brent, O.P.

---

10. For example, Wis. 13:1–9, Ps. 19:1–4, Sir. 42:15–43:33.
11. Ex. 3:14.

CHAPTER 7

The Nature of Creation

The purpose of this chapter is to present the Church's understanding of creation, and to do so in light of the wisdom of St. Thomas Aquinas. The two arguments for the existence of God offered in the previous chapter, the argument from world order and the argument from contingency form the background for penetrating the mystery of creation.

It seems that when most people think of creation, they think of it as a past fact, a historical fact. To the minds of many, if the world was created at all, it was created in the past, and God's work of creating it is now over and done. Let us call this a historical account of creation.

Thomas Aquinas does *not* hold a historical account of creation, and he does not think of creation merely as a past fact that is now over and done. Rather, he holds a metaphysical account of creation. To say that God creates the world is to say that God gives being to contingent things: "to create, properly speaking, is to cause or produce the being (*esse*) of things."[1] To be a creature is to depend on God for the very act of existence. Creation is the emanation from God of everything other than God, and this emanation is a present reality.[2]

An analogy helps to understand. Just as the sun shining in the sky illuminates the atmosphere, so God gives being to everything other than God. The analogy illustrates two critical points. First, just as on a sunny day the sun is *now* causing the atmosphere to be illuminated, so God is *now* causing contingent beings to be. Second, just as on a sunny day the illuminated atmosphere is *now* dependent on the sun for its illumination, so contingent beings are *now* depending on God for their very being. The relation of creature as a creature to the Creator as Creator is thus primarily present and not past, primarily vertical and not horizontal. So long as God now gives being to contingent things, and so long as contingent things have this present vertical dependency on God for their very existence, then God creates them and they are God's creation.

Since this metaphysical feature of depending on God for being is primarily a present and vertical feature of things, it is compatible with a variety of historical possibilities about how things have unfolded or developed in history from horizontal causes or preceding conditions. Aquinas even considers the possibility of a world history such that the world of nature has existed without beginning, and he insists that even if the world had never began to exist it would still be God's creation. This point is difficult to grasp, but worth considering with care.

Let us assume, for the sake of a thought experiment, an everlasting sunny day. By definition, such a day is without beginning and without end. On such an everlasting day, the illumination of the atmosphere would be everlastingly dependent on the sunshine. Similarly, even if the world of contingent beings were everlasting, it would be everlastingly dependent on God and God

---

1.   *ST* I, q. 45, a. 6.
2.   *ST* I, q. 45, a. 1.

would be giving it being now and always. Each contingent being would presently depend on God for its very act of existing, and hence it would be God's creation.

This metaphysical account of creation may be illustrated in another way. Let us diagram the world of nature (or more broadly the order of contingent beings) as follows:

GOD

DIAGRAM I

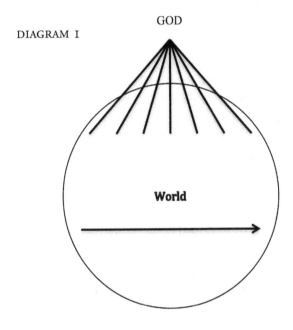

World

In Diagram 1, the rays coming down from the dot represent the emanation of existence to each contingent being in the world, the circle is the world of Nature (or the order of contingent beings), the area of the circle is space, and the arrow in the circle represents time. Just as God is beyond the circle, so God exists beyond the world. God exists beyond time and beyond space. God does not create the world in time, but with time—or having time in it. God does not create the world in space, but creates space in the world—or a world full of space. God gives being to the whole world and all things in it, including time and space, from out of his very Being. (And since his very Being is also Eternity, one could say that God creates all things in time from Eternity.) This diagram shows us, we can say, the essence of creation.

The diagram helps us to arrive at the critical insight that without changing the diagram, that is, without contradicting the essence of creation, we can conceive of a number of possible historical unfoldings of things in the world. For example, the timeline in the above diagram has beginning at the far left to represent a beginning of time. But if the line were changed to have an arrow at the left, representing that time did not begin or end, the overall relation of God to the world would remain the same. A universe of everlasting time would still be a created one. Diagram 2 shows how it would look.

DIAGRAM 2   GOD

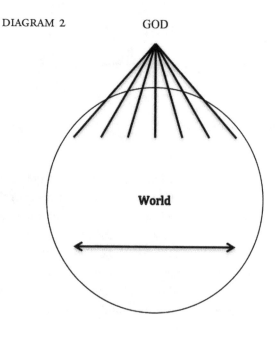

A world of time without beginning or end would be a creation of God because the world and all things in it, including space and time, emanate from God and depend on God for their very existence.

Setting aside the issue of whether time began or not, we could also make the diagram more specific regarding chemical or biological facts.

If we were to say that in the world there are a fixed number of biological species (A, B, C, etc.) that have never evolved, but remain forever as the same species, then we would simply put "species remain the same" on the time line. Diagram 3 shows how the world would look.

DIAGRAM 3

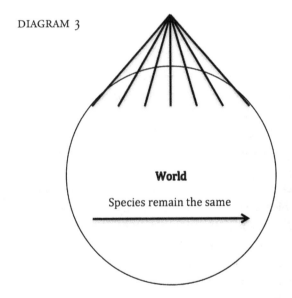

World

Species remain the same

If, however, we say that from an original species or set of species still others evolved by random mutation and natural selection, then we simply put "species evolve" on the line.

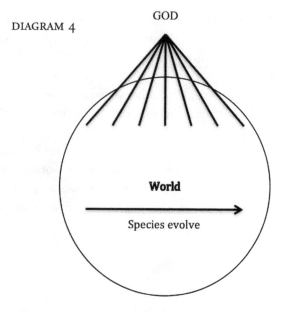

Whether the world was comprised of a fixed number of bio-logical species that remain ever the same or an evolving number and diversity of species, the world is created. For in either sce-nario, each thing in the world, including space and time them-selves, would emanate from God and depend on him for their very being.

In fact, there are a variety of possible scenarios about the ways in which the elements, molecules, and life-forms may have developed, evolved, or remain fixed, and in all these possible sce-narios the world and each thing in it would be created so long as it was presently depending for its very being on God or on the emanation of things from the eternal God.

Two further important and related points are in order. First, God creates the world freely. He does not have to give being to things, but he freely wills to do so. Why does he freely will to do so? Why does God create the world? In other words, what is the point of creation? We come to the question of the ultimate purpose of the world, and this question touches on God's wisdom in creation.

Second, God creates the world wisely. According to the argument from world order, God is the source of the order, coordination, and harmony of Nature. When we contemplate that order, we can see in it a plan or intelligibility. Biology and ecology both reveal a particularly breathtaking beauty, harmony, and order in things. We can see *that* there is an intelligible plan for the world, and we can see *what* the plan is to some extent (at least to the extent of knowing something of the general order of nature), even though we cannot discover God's plan for each particular thing in detail.

When we study the order of the world, we find a distinction between *persons* and *things*. Persons are beings capable of knowledge and love, things are beings without the capacity for knowledge and love.

To the question of what is God's ultimate purpose in creating the world, the answer of Aquinas (in agreement with all the theologians of his day), is that God creates the world so that created persons may contemplate the order of things and by doing so come to know and love God to some extent—at least to the extent that his attributes are distantly reflected in the order, beauty, and harmony of things. In other words, God creates the world so that he may be known and loved by created persons. Another way to put this is that God created the world for his glory, i.e., to show his attributes to us through the world of nature. Nature is God's sign language to us. Nature is the first revelation of God to human beings, a revelation accessible through natural reason apart from faith, and nature is the background for a second revelation of more intimate secrets of God: his Triune life, his plan of salvation through the Incarnation and

Paschal Mystery, his call to eternal life in communion with God and the saints in the beatific vision. The second revelation calls for faith in God revealing Himself in a higher way to us than by nature and reason.

The doctrine of creation is precisely the denial of the view that God is hidden or inaccessible. The doctrine is the announcement that God creates the world in order to come out of hiding and be present, in different ways, to both the reason and faith of persons.

James Brent, O.P.

❧ CHAPTER 8

God's Knowledge and Love in Creation

Necessity is the mother of invention, or so the saying goes. Human ingenuity is at its best when it sees a dire need, draws on all its available resources, mental and physical, and finds a way to fill that need. While this is surely the case in many successful inventions, it is amazing how many things there are in our daily life that we simply take for granted that were not sought out but were stumbled upon either by accident or even by error. Superglue and Teflon, plastic and vulcanized rubber, corn flakes and chocolate chip cookies, stainless steel and the microwave oven were all inventions that were stumbled upon rather than sought out. Both necessity and chance lead to invention.

Looking back over the history of ideas, numerous religions and many thinkers have tried to attribute the origins of everything either to necessity or to chance. There is a whole range of creation myths in ancient cultures that see creation either as a necessary byproduct that results from the interactions among various deities—for example as something that is born biologically from the gods—or as an accidental result of some activity associated with the gods—for example as the reside of a battle or a sacrificial offering. The Babylonian creation myth called the Enuma elish (7th century B.C.), for example, declares that the earth and the skies were fashioned by the Babylonian god, Marduk, from the corpse of another god named Tiamat. In contrast, many philosophers of the ancient world rejected such a contingent beginning for the world and sought to ground creation in necessity. In their view, the origins of the universe can be linked to a necessary emergence from a chaotic beginning or to a necessary procession from a primordial divine being.

A Judeo-Christian understanding of creation has always rejected both of these extremes, emphasizing the intentional and free nature of God's creativity. In creating out of nothing, God knew what he did, did it freely, and did it well. Importantly, this divine doing is not restricted to that moment "in the beginning." Rather, the whole of God's creative activity as he sustains and orders all of creation throughout time is rooted in his intelligence and in his love. St. Thomas Aquinas explains:

> God has brought things into existence not through any necessity of his nature but by his will.[...] God is infinite in power. Consequently he is not determined to this or that effect, but is undetermined with regard to all effects. [...] Hence effects proceed from God according to the determination of his will. And so he acts, not by a necessity of his nature, but by his will. This is why the Catholic faith calls the omnipotent God not only "Creator," but also "Maker."[1]

For Aquinas, God was radically free to create or not to create, and he created everything as ordered to himself.

---

1.   *Compendium of Theology*, no. 96.

While it surely takes a good dose of intelligence to recognize when a mistaken byproduct has some hidden potential, properly speaking the fortuitous accident is not an intelligent act. To act through intelligence means to foresee the action and the result in our mind and then to bring it about in reality. This is best seen by analogy to the artist who envisions an image in his mind and who then applies all his talent and skill to bring about that image in a painting or in a sculpture. This is the sort of activity we attribute to God in creation. He foresaw an image, an exemplar, of each and every created thing in his divine intellect and set about to bring each thing into reality. Aquinas compares the ideas that God uses to create the world to the forms in the mind of a builder:

> In other agents (the form of the thing to be made pre-exists) according to intelligible being, as in those that act by the intellect; and thus the likeness of a house pre-exists in the mind of the builder. And this may be called the idea of the house, since the builder intends to build his house like to the form conceived in his mind. As then the world was not made by chance, but by God acting by his intellect, there must exist in the divine mind a form to the likeness of which the world was made.[2]

Of course, as with any analogy of God with his creatures, this comparison is an imperfect comparison. However, it is worth seeing exactly how it is imperfect.

For every undeniable artistic masterpiece, there are plenty of nice tries and flat out mistakes that are produced. It is almost impossible for the finished product to live up to the image the

---

2. *ST* I, q. 15, a. 1.

artist intends, and this can be for either of two reasons. First, there is the limitation of the artist. Try as we might, however beautiful a scene we might imagine, most of us if armed with canvas and paintbrush would end up with something more fitting for the refrigerator or dumpster than a museum. We simply do not have the technical skill and aptitude to bring about what we see in our mind, and though we can certainly get better with practice and training there will always be some gap, perhaps slight, between our ideal and the reality.

On the other hand, there can be a limitation in the materials. Even an artistic genius, if handed a crumpled up piece of brown construction paper and a half empty box of crayons can only do so much. Sure, the result would most likely be amazing beyond anything the rest of us could hope to achieve, but the imperfection of the materials would prevent the perfect realization of the mental image of the artist, or at least limit the possible images he could try to realize.

Aquinas makes clear that in God neither of these limitations applies. While the human artist has finite power to bring about the desired image in reality, God is omnipotent and so there is no lack on his part that could limit his creative activity.[3] Further, no limitation in the material of creation can limit his creative activity because no material is outside his creative power.[4] Thus, whatever image God sees in his divine mind he can make a reality. The question then remains, if God has the power to create whatever he can think of, why did he make the particular things that exist, including you and me?

When we look at the world around us and at ourselves and see imperfection and even evil, it can be hard to make sense of

---

3. *ST* I, q. 19, a. 6.
4. *ST* I, q. 45, a. 2.

the idea that God is both completely good and all-powerful. One way that well-meaning thinkers have tried to reconcile these ideas is to posit that, in some way, God had no choice in the matter. This general idea often comes with the idea that this world, despite all its lumps and bumps, is the best possible world that God could have created. The claim is that, while there are individual problems and evils, if any of them were removed the overall effect would actually make things less perfect. Since God is good, the argument goes, he would have made a perfect world, and this world is that world. While this idea exonerates God of all the individual blemishes we cannot help but see, it also guts his omnipotence and freedom in creating and distances him from any particular piece of his creation, including each one of us.

If this is the best possible world, God did not create you simply because he wanted you to exist. Rather he created you, even if he wanted you, because you were necessary to make things work out better for the whole. While there is an honor in being a small cog in a beautiful machine, God's will to create each of us is more than that. Rather than claim that this is the best possible world, Aquinas affirms that, although this world is good because God made it, it is not some absolute of goodness.[5] God alone is truly, fully good, and he had no need to create anything at all. his goodness, in and of itself, would be no less if he never created anything, and is not increased because he did create. God was absolutely free to create whatever he wanted to create, and he created us. It is true that there is a greater goodness in the whole than in any particular part and that there are individual perfections that are ordered to the greater good of creation as a whole, but these facts do not constrain God's freedom. In his intellect he sees every possible created thing and every possible creation, and

5.   *ST* I, q. 25, a. 6.

He chose to make this one real and, in so doing, chose to make each and every part of it exactly as it is.

This freedom derives from God's will and his ability to choose to create whatever he wants, but the first mover of the will, the root of every choice, is love.[6] When we look at the imperfection in the world, sometimes we can see how it is ordered to some greater good and we can perhaps even understand why God might have chosen to make things in this particular way, but often it is hard to see how goodness might overcome some specific evil. We will touch on the idea of God's providence, his overarching plan for creation, in a later discussion but there are two truths we can be certain of. God understood exactly what he was doing when he created each and every aspect of reality, and everything that exists does so because he loves it, including each and every one of us.

Thomas Davenport, O.P.

---

6.   *ST* I, q. 20, a. 1.

❧ CHAPTER 9

God's Providential Governance of Creation

The experience of watching a construction site can be quite engrossing. There are dozens of people, many using amazing tools and machines, each with its own function and each working on a different portion of the site. At times the workers can seem completely independent and almost oblivious of one another, as if the site were a chaotic jungle of random jobs. Over time though, the coordination of these individual efforts becomes apparent, as one worker's efforts prepare the way for that of another, whose work links up with a third. The seemingly haphazard collection of skills and tasks comes together into an ordered structure with a clear design and purpose. Of course, this emergence of order is no happy accident. Much time and effort went into the planning of the project, laying out every minute detail in diagrams and blueprints. Further, someone was responsible for knowing and understanding those blueprints and ensuring that all of those disparate workers stayed on task and contributed to the greater order of the building as a whole, namely the foreman.

In many ways, the foreman is a helpful analogy for understanding God's providence. Of course the analogy limps in places, and it will be helpful to clarify what truths about God it illustrates and those that it cannot. Divine providence refers to God's pre-existing knowledge of every aspect of creation, somewhat like the foreman's knowledge of the blueprints for the building. Just as the foreman knows about and is responsible for the whole building in its entirety, not in an abstract and general way, but down to very precise details and particulars, God's providence is over the entirety of creation, covering all places and times. It is not abstract and general. Aquinas insists that God's providence covers every detail of every created thing, without exception: "But the causality of God, Who is the first agent, extends to all being, not only as to constituent principles of species, but also as to the individualizing principles; not only of things incorruptible, but also of things corruptible. Hence all things that exist in whatsoever manner are necessarily directed by God towards some end."[1]

Further, the foreman knows the reason, the goal, behind each and every task related to the construction of the building. He knows that the space must be dug for the foundation in a very particular way so that the foundation can be laid correctly to properly hold the supporting structure which will give strength to all of the internal floors and walls and so forth. God's providence also extends to goals and ends. God knows the proper end of every part of creation, for he made each of them, and directs each of them to that end. Furthermore, he orders all of these lesser ends, one to another, for the greater glory of the whole of creation and, ultimately, to the last and greatest end, God Himself.

As marvelous as the blueprint or plan may be, it is not meant simply to be thought about and admired by the foreman, but to be instantiated as a real building. This distinction between the plan as conceived and the execution of the plan is, for Aquinas,

---

1. *ST* I, q. 22, a. 2.

the distinction between two aspects of God's providential ordering, providence, properly speaking, and government:

> Two things belong to providence–namely, the type [idea] of the order of things foreordained towards an end; and the execution of this order, which is called government. As regards the first of these, God has immediate providence over everything, because he has in his intellect the types [ideas] of everything, even the smallest; and whatsoever causes he assigns to certain effects, he gives them the power to produce those effects. Whence it must be that he has beforehand the type [idea] of those effects in his mind. As to the second, there are certain intermediaries of God's providence; for he governs things inferior by superior, not on account of any defect in his power, but by reason of the abundance of his goodness; so that the dignity of causality is imparted even to creatures.[2]

Just as the foreman assigns laborers to particular tasks on the construction site, even though they may not understand the whole of the project, God deigns to give creatures a particular role in the government of his creation to bring to fruition some part of his larger plan.

Here, though, is one place where the weakness of the analogy arises. Even though the foreman may know how to work every machine and fulfill every task needed, he needs the help of the laborers to complete the task in a reasonable time or, really, at all because of the limitations of his own power. Further, because of the trust he places in his laborers and their own limitations, part of his responsibility is to react to problems that may arise in

---

2. *ST* I q. 22, a. 3.

their completing each task and adjust the work prudently to best approximate the original plan.

God does not rely on creatures to execute his providential order of creation because of any lack or weakness on his part. Rather he invites them to cooperate with his providence because of the abundance of his goodness. He desires to impart the dignity of causality, the dignity of having a full and true role in the greater order of nature, to his creatures. As we discussed earlier, God's primary causality reaches down to every aspect of creation, but he chooses to work through the secondary causality of created instruments. Further, the providence of God is not reactive like the governance of the foreman. God is never surprised or caught off guard by the limitations of his creatures. Even their imperfection, their fallibility, does not escape his knowledge. Because the flow of time is a property of creation itself, God is not subject to our moment by moment experience, but has full knowledge of every moment, past, present and future, in one eternal now.[3] His providential knowledge of what is for us the future, is unchangeable and certain because, to Him, it is already present.

Note that these two claims, the true causality of creatures in the created order and God's certain and unchanging foreknowledge, may seem to be in tension, even to be in contradiction. If God already knows what will happen, is there really any such thing as a chance occurrence or an accident? More importantly, if God already knows what every choice will be, is there any such thing as a free choice? We will look at this tension more closely in the next few essays. However, following Aquinas, we can say that the short answer is, yes: There is chance and there is free will even though God's providential knowledge and his governance of creation are exhaustive.[4]

---

3. *ST* I, q. 14, a. 13.

In sum, divine providence is simply the extension of God's intimate knowledge of each created thing to the totality of creation, without sacrificing any of its depth. This providence is both comprehensive and extremely particular. God was not in any way constrained in what he could have created and so his choice of this particular providential order was made freely, a choice made purely from his divine wisdom and love. Thus, when we catch a glimpse of the beauty of the order of the created world, in nature itself or in human activity, we can take delight in the fact that we are catching a glimpse of God's providence. Even when that order escapes us, and the tragedy of evil stares us in the face, we can take some comfort in the knowledge that nothing escapes God's wise and loving providence and that, ultimately, all things are ordered to the good, that is, to God Himself.

Thomas Davenport, O.P.

---

4. See Chapters 10, 11, and 12.

CHAPTER 10

Randomness, Chance, and the Providence of God

What exactly is randomness? We could begin by looking to a standard definition, such as "having no specific pattern, purpose or objective." Further, we could imagine some example of randomness, like a coin flip, a die roll, or a roulette wheel spin. If we ponder it still further, randomness might seem to suggest, not simply unpredictability, but chaos and unintelligibility. How can we reconcile the apparent lack of intelligibility and order in the random and chance events of history, both in the evolutionary history of life and in the history of human civilization, with a divine providence that knows and orders all things?

In order to answer this question, we need to understand what exactly makes random events unpredictable. We need look no further than a surprise birthday party to see that what is unpredictable for one person need not be so for a whole host of other people. The paradigmatic example that Aquinas inherits from Aristotle is the case of the meeting of two servants in the marketplace. It seems like a chance circumstance to each of them, but is a completely predictable and planned event for the master who intentionally sent them both to the market, without their knowledge of the fact.[1] Randomness can arise from two, or more, perfectly natural processes each working towards some natural end, that end up interacting in a way that interrupts them both. For instance, it is natural for a wolf to chase its prey and natural for a rock loosened from a cliff to fall. However, it would be a random or chance event if this particular wolf happened to be struck down and killed by this particular rock, mid-chase.

Even our ideal cases of games of chance are only random because the circumstances surrounding them, the unknown initial conditions, make them difficult to predict. There is nothing particularly mysterious about the physical processes involved in the roll of a die. If the initial conditions were well controlled or carefully examined the randomness could be lessened or even eliminated. Gamblers have beaten the odds in roulette using computers to work out the physics of each spin, and engineers have made a mechanical coin flipper that uses the same initial conditions to always come up heads. Newtonian physics attempts to extend this simple analysis to all of nature. At the extreme, the argument made by some physicists was that if we could know the exact conditions of every part of the universe at any one moment, we could use the laws of nature to calculate the exact conditions of every part at any future moment.[2] While

---

1.   *ST* I, q. 22, a. 2 ad 2

2.   Pierre-Simon Laplace famous stated this idea as: "We ought to regard the present state of the universe as the effect of its antecedent state and as the cause of the state that is to follow. An intelligence knowing all the forces acting in nature at a given instant, as well as the momentary positions of all things in the universe, would be able to comprehend in one single formula the motions of the largest bodies as well as the lightest atoms in the world, provided that its intellect were sufficiently powerful to subject all data to analysis; to it nothing would be uncertain, the future as well as the past would be present to its eyes. The perfection that the human mind has been able to give to astronomy affords but a feeble outline of such an intelligence." *Essai Philosophique sur les Probabilités*, forming the introduction to his, *Théorie Analytique des Probabilités* (Paris: V Courcier); repr. F.W. Truscott and F.L. Emory (trans.), *A Philosophical Essay on Probabilities* (New York: Dover, 1951).

this task is obviously practically impossible, the principle is that the world is ultimately deterministic where any randomness we encounter is only apparent, arising from a lack of full knowledge on our part.

Since the advent of quantum mechanics in the early twentieth century, this deterministic view of physics has fallen on hard times. On account of many unexpected results, quantum mechanics gave rise to the Heisenberg uncertainty principle, which put a hard limit on just how exact our exact picture of any object, let alone the universe, could be. This was not simply a statement of some technical difficulty in measuring the properties of an object, but a fundamental limit on how precisely different properties of an object, for instance its position and its speed, could be known. This result and its apparent placement of chance at the core of all phenomena bothered many physicists, including Albert Einstein, who summed up his dissatisfaction with this loss of determinism in his oft-quoted phrase, "God does not throw dice."[3] While the success of quantum mechanics has led most physicists to accept the loss of the tidy picture of classical determinism, there are still deep disagreements about the nature and source of the apparent randomness in quantum mechanics, including attempts to revive determinism. This is not the venue to try and settle this problem, but we do need to consider the impact of our current understanding of randomness and of determinism on our understanding of divine providence.

It is tempting to think that divine providence requires something like determinism. If for no other reason, it is easier to see

---

3. Einstein used variations of this quote numerous times. An early example is from "Letter to Max Born (4 December 1926)," in *The Born–Einstein Letters*, trans. Irene Born (New York: Walker and Company, 1971).

how divine providence would work if everything that happens follows a fixed set of mathematical laws. In this case, God could simply be a good physicist who knows both the state of the universe and the laws of nature he has designed according to which things happen. In this view, God is infinitely more powerful and knowledgeable than any human physicist, but his providence would basically be just a matter of solving a very hard physics problem. However, in addition to the difficulties that this requirement of rigid determinism presents for free will, this link of providence to physics faces the problem that physics itself has complicated our picture of nature. If there really is a fundamental limit on how precisely any physical system can be known, then how can divine providence be both comprehensive and infinitely particular, down to the last detail? If certain phenomena are actually inherently unpredictable, how could God actually know the future? The fact that we are able to make very accurate predictions based on the probable outcome of events would still not allow us to explain how God could providentially know and guide his creation so that it is always ordered towards the good.

The answer, of course, is that we cannot bind God's divine knowledge and power to the limitations of created knowledge and power. This is not a physical claim about quantum mechanics or any particular phenomenon, nor is it a simple appeal to mystery. It is a philosophical claim about the way that God creates and sustains the universe. God's providence is not a well-executed mathematical calculation, but an omniscient and omnipotent mastery of all reality that brought the universe into existence out of nothing and sustains it in existence at every moment of time. God chooses to grant his creatures the power to cause. He creates things not just to happen, but to happen in a particular way. Aquinas is clear that "whatsoever divine providence ordains to happen infallibly and of necessity happens infallibly and of necessity; and that happens from contingency, which the plan of divine providence conceives to happen from contingency."[4] The fact that a particular phenomenon has an element of randomness or contingency does not remove it from divine providence. God's creative power is such that the very powers that allow a creature to act and to cause, even to cause contingently and by chance, depend at every moment on his sustaining power. Whatever happens in the world, whether it is a radioactive decay, a biological mutation, a decision to sin, or a decision to praise Him, does not catch God by surprise. In fact, he gives his creatures their existence and their natures that allow them to decay, to mutate, to sin, or to praise. This type of knowledge seems to go against our very understanding of what knowledge and causation are, but that is because we are only familiar with how created causes know and work. God is not another part of nature. He is not even the greatest part of nature. Rather he is nature's author and sustainer. He is the Creator, totally other to the created universe.

Thomas Davenport, O.P.

---

4.   *ST* I, q. 22, a. 4, ad 1.

❧ CHAPTER 11

Providence and Freedom

Time travel is a tricky business. While the prospect of seeing the future or reliving some historical event is extremely enticing, there are a whole host of dangers and difficulties including the prospect of getting stuck in a loop, reliving the same events over and over until you get them right. Perhaps you accidentally intervened in some past event that has corrupted the present and you feel the need to repeatedly go back and try to fix things. Or maybe the repeated time travel is out of your hands and you have to tread a perfect path to break out of the cycle. It is one of those undeniable truths of science fiction that it takes extremely careful planning and coordination, along with a good bit of trial and error, to manipulate the circumstances and make sure everyone else acts in exactly the way you need them to in order to break out of a time travel loop.

As odd and entertaining as this fictional conceit can be, it is not so far from the way that some have tried to understand the relationship between the all-knowing providence of God and the freedom of individual human wills. But before we delve into time travel and omnipotence, we should first retrace a few steps. In a previous chapter, we described God's providence as the intimate knowledge of every aspect of creation down to the smallest detail, and adverted to the fact that, for Aquinas, this foreknowledge does not conflict with human freedom. We have already touched on some difficulties with this claim, but they deserve a closer treatment.

The first difficulty is this: If God knows what choices I will make in the future, how can I be free to make those choices? To answer this question, we need to recall that God's knowledge and action isn't like the knowledge and action of some creature. God knows everything about us, but not in the way that Bob, our nosy neighbor, knows everything about us.

When Bob watches us cut our lawn, he knows the infallible truth that we are cutting our lawn right now. This knowledge does not impinge upon our freedom because it is simply a statement of what we have freely chosen to do right now. If Bob tries to predict what we will be doing at 9:30 on Saturday morning he may be able to guess with some plausibility, particularly because Bob is nosy and knows that we have a habit of cutting our lawn at that time every week, but his knowledge of our future is contingent. Specifically, it is contingent upon our free choice in the matter, as well as upon all sorts of other unpredictable circumstances that might influence that choice.

If God has infallible knowledge of our future, then it may seem like he must have some secret knowledge that makes contingent things not really contingent but necessary, and our free choices not really free but already chosen for us. Aquinas insists that God's infallible knowledge of our future is not some secret predictive power because there is no such thing as the future for God. Rather, in his transcendent eternity, which is outside the flow of time, all events from any time are present to him in one eternal now. Thus God's infallible knowledge of our future free acts is less like Bob's guess about our whereabouts on Saturday morning and more like what he knows while peeping over the fence at us in the middle of our freely chosen act of cutting our lawn. "Hence it is manifest that contingent things are infallibly known by God, inasmuch as they are subject to the divine sight in their presentiality; yet they are future contingent things in relation to their own causes."[1]

While this may help us understand how God's knowledge is reconcilable with human freedom, there is still something problematic with thinking about God as nothing more than a very observant neighbor. In his wisdom, God also orders all things as part of his loving plan for creation. This involves not only knowing what will happen but also directing it to come about in a certain way. God works through his creatures in this act of governance, giving them true roles of causality, but never in such a way that anything is beyond his reach or power.

Here is where a second difficulty with providence and free will arises: If God's providence is more than just knowing all things, but involves the execution of a particular plan for creation, how can our free actions be part of this plan and still truly remain free? Some have argued that for human freedom to truly remain free, it must, in some way, be off limits from the direct causal action of God. To get around this problem they propose that God never acts directly to move us to a particular end, but knows exactly how we would freely act in any possible circumstance and so orders the circumstances to bring about the particular free act that fits into his plan. While God is clearly not stuck in one of those time travel loops of science fiction, on this view, he is still constrained in a similar way to indirectly manipulate human action, not based on trial and error but on his infinite knowledge of how we would act in different situations, given the chance.

---

1. *ST* I, q. 14, a. 3.

Aquinas is much more direct in how he sees God ordering man's free actions as part of his providential plan. God is the creator of human nature and the creator of our rational souls, so we are utterly dependent on him for our existence and especially for our ability to think and to choose. "God moves man's will, as the Universal Mover, to the universal object of the will, which is the good. And without this universal motion, man cannot will anything. But man determines himself by his reason to will this or that, which is a true or apparent good."[2] God's Divine plan includes not only the fact that certain things happen, but that they happen according to the nature of the creatures involved, which, in the case of man, includes self-determination.[3] The results of this self-determination never catch God off guard and would not be possible if he did not actively empower us to choose.

The fact that God's infallible knowledge of our actions and his involvement in our very act of willing does not destroy freedom can seem contradictory. For us, certain knowledge only comes by removing any and all contingency, and inducing others to act always involves a certain coercion or violence. God does not act like us, though. His infallible knowledge and omnipotent will are gentle enough to preserve true freedom on our part such that, even though we are utterly dependent upon him for our ability to act, we are truly responsible for the results. Even here, though, God does not leave us alone to the whim of our fallible desires, for by the mystery of grace, which would take us far beyond the scope of this chapter, he can help us to choose the true and highest good, namely Himself.

Thomas Davenport, O.P.

---

2. *ST* I-II, q. 9, a. 6.
3. *ST* I-II, q. 10, a. 4.

❧ CHAPTER 12

Divine Providence and the Mystery of Evil

It is an undeniable fact that we have a greater understanding of nature and better techniques for harnessing the power of nature today than in previous centuries. From this privileged position we can sometimes think that many of the questions and difficulties this new understanding raises, whether moral, philosophical, or theological, are unique to our present age. They must be questions that previous ages could not have fathomed with their limited understanding of the world. It can be a surprising and happy discovery to find that we, as a society, are not so isolated and alone in our struggles, and that many of these difficulties are but appearances of age old questions.

Consider the apparent evil of extinction that occurs during evolution. A society that thought all living things had existed exactly as they are from the beginning of the world could hide itself from the fact that whole classes of animals and plants were biological dead ends, left on the cutting room floor of history. From the early days of thought on biological evolution many saw Tennyson's "nature, red in tooth and claw"[1] as a pithy summary of the violent and uncaring way that natural selection suggests species continually try to overpower their competitors. How could God's wise and loving plan include such widespread pain and death? Doesn't the central role that evil plays in the evolution of life contradict the existence of divine providence or, even worse, the goodness of divine providence? Is this not a new difficulty that our civilization has never had to confront before?

Perhaps, if you have been reading through these chapters, you will not be surprised to hear that the answer to these last two questions is "no." We do not need the theory of evolution to tell us that the natural world is violent and bloody. Simple observation of the animal kingdom reveals that precious few animals, of any species, die at peace after a long fulfilling life. Obviously, carnivores only survive by the death of their prey but, further, no predator is without a predator of its own and even the most dominant predator is only safe until a close relative, perhaps even its own offspring, is strong enough to overpower it. In fact, our pre-modern ancestors[2] were almost certainly more aware, by personal experience, of the violence of nature than those of us who only interact with wild animals in zoos, nature documentaries and "snake vs. crocodile" YouTube videos.

Unsurprisingly, Aquinas takes up the question of evil in nature in his treatment of divine providence. In holding that God's providence extends to every detail of existence he insists that this universal care does not exclude the possibility of particular defects and evils:

> Since God, then, provides universally for all being, it belongs to his providence to permit certain defects in particular effects, that the perfect good of the universe may not be hindered, for if all evil were prevented, much good would

---

1. "In Memoriam A.H.H."
2. Aristotle, *History of Animals*, 9.1.

be absent from the universe. A lion would cease to live, if there were no slaying of animals; and there would be no patience of martyrs if there were no tyrannical persecution.[3]

His first example of an evil that God allows, in this case an example of physical evil, is of violence in the animal kingdom which, if completely prevented, would eliminate the lion and all other carnivores, limiting the expression of God's creative power. This is not to say that God delights in or directly wills suffering for the sake of suffering, but that he allows it for a greater manifestation of goodness in creation, a greater manifestation of his goodness.

The other example Aquinas discusses is one of sin or moral evil, the persecution of the martyrs, which might cause one to wonder if the physical evil of animal violence is really just a manifestation of sin, something that was foreign to God's original plan but that he allows for a time in order to redeem it in the future. In fact, some Fathers of the Church thought exactly this, seeing the bloodiness of nature as a defect resulting from the fall of Adam and claiming that all creatures ate plants in Paradise. Setting aside the question of plant genocide, it is interesting to see that Aquinas rejects this view, arguing that moral evil has many negative effects, including corrupting human nature, but it does not destroy our nature or the nature of other things. Sin is not powerful enough to change what an animal eats.[4]

The distinction between physical evil and moral evil is important for understanding God's causative role in our imperfect world. First and foremost, God, as goodness itself, primarily and directly wills the good in all cases. Second, certain particular physical evils, although not willed directly, can be said to be

---

3. *ST* I, q. 22, a. 2, ad 2.
4. *ST* I, q. 96, a. 1, ad 1; *ST* I-II, a. 85.

caused by God as part of his wise ordering of the whole. Thus the physical evil of violence in the animal kingdom, even though it introduces particular evils, is directly part of God's divine plan for the world and makes room for the greater goodness of the whole of nature. When God willed to create lions as carnivores, he also had to permit their killing of antelopes. Finally, the willed evil of sin, a moral evil, by which we freely choose to act and order our hearts against God's design and our own good, is not caused by God at all but only by our own free choice. Thus, while God holds us in existence and gives us the power to choose, when we choose to sin we are acting, by definition, against his will. This is not to say that his providence is in any way frustrated by our sin or that he is caught off guard by our actions or that he cannot bring great good out of our malice, but that the cause of our sinful action is ultimately our own free will.

The problem of evil and sin in the world has vexed men for millennia and the Church has had to address it in many guises from the very beginning of her existence. Despite what can seem like evidence to the contrary, she has confidently proclaimed God's divine providence over all aspects of creation, without claiming to always understand how God's goodness ultimately triumphs over any particular evil, whether physical or moral, that we face. New experiences and new discoveries will surely raise this difficult question again and again in new semblance. Many of these, like the evil of extinction (which will be looked at more closely in a later essay), may be understandable when considered from the proper perspective. Ultimately, though, the Church's final answer to these problems, whether understandable or incomprehensible, is that Christ has won the victory over sin and death and that God wills to invite his whole creation to participate in that victory.

Thomas Davenport, O.P.

CHAPTER 13

The Authority, the Character, and the Interpretation of Sacred Scripture

In the next six chapters, which deal with revealed theology, we will examine the compatibility of the theory of evolution with the biblical accounts of creation. In this chapter, we will establish the authority of the Bible, its mediated character, and its authentic interpreter.

The Bible is called the "word of God" in that God is the primary author of the Bible. The human authors of sacred Scripture, such as Isaiah or Matthew, are also true authors of the text, but they are secondary, not primary, authors. How so? They are true authors in that they wrote what *they* understood, and they wrote it using their own intellectual and physical powers. They are secondary authors, and God is the primary author, because God used these human authors as instruments in producing this written text—as one uses a pen as an instrument in writing a note.

As an illustration, when Albert Einstein wrote out the equation $E = mc^2$, he—and not the pen he was writing with—was given the credit for producing that formula. He is the author of that formula, and the pen was the mere instrument with which he wrote it. The pen was not capable of producing the formula, but only of placing ink efficiently on paper.

Now, in the case of sacred Scripture, the human authors were capable of writing a text in a human language that they knew, but only God is able to make this an authoritative and inspired text that brings sinners to repentance and to eternal life. God's power is so great that he is able to use sinful human beings—despite their cultural biases, personal fears, and other limitations—as instruments in writing out his saving word.

Thus, sacred Scripture is *mediated*: The Bible has true secondary authors. It is not a simple dictation by God, where the human author falls into a trance and writes what he himself might not have understood or agreed with. Even so, sacred Scripture is *inspired*, because God is the primary author. It is insufficient to say that the human authors were inspired by a religious experience and then wrote it down. Surely they were, but what makes the Bible unique among all written works ever produced or yet to be produced, is that God is the primary author of this book, and of this book alone.

This mediated character of sacred Scripture explains why the Bible can be translated into other languages without losing its status as the inspired word of God. Christians affirm that the Word became *flesh*, not a book in Hebrew or in Greek. By contrast, when Muslims translate the Quran, it is not clear whether this is the true Quran because, as they understand it, God's original tablet in heaven is in Arabic, and no translation renders the original perfectly.

This mediation of the Scriptures through the community of believers is inescapable and should not be viewed as replacing God's action as the primary author. For example, Jesus spoke in Aramaic, but the New Testament is recorded in Greek—which would leave us one step removed from Christ, if it is not the Holy Spirit Himself who worked through this mediation. Similarly, the determination of what books belong in the Bible, which manuscripts are legitimate, and when no more books may be added, was made by the Church, under the guidance of the Holy Spirit. Furthermore, valid translations of the Bible—on which the great majority of Christians rely since they cannot read ancient Greek and Hebrew—have to be approved by a legitimate ecclesiastical authority, again under the guidance of the Holy Spirit.

Now, the Bible is privileged above all other writings, be they authoritative doctrinal declarations of Church councils, mystical writings of the greatest saints, or brilliantly penetrating and universally integrative theological treatises. This is important to establish at the outset, because if the Bible rules evolution out, then for the Christian, the argument is over. There is no higher authority to which one can appeal.

In addition to having the highest authority, Scripture is of the greatest importance because it is God's revelation of Himself and of our own ultimate vocation. The Bible teaches us things about God and ourselves that we could never discover on our own, such as the Trinity and the Resurrection. It also confirms things that human society could figure out on its own but only with great difficulty, such as the Ten Commandments.

As Aquinas explains, God reveals even these truths accessible to human reason because only the brightest minds over hundreds of years could reason these things out, and their conclusions would still include many errors.[1] This aspect of Scripture bears significantly on the question of evolution, because we can expect that Scripture has something to teach us about our first origin and our final end.

An inevitable question concerning a right understanding of the Bible is: Who has the authority to interpret Scripture? A significant difficulty that attends written revelation is the possibility

---

1. See *ST* I, q. 1, a. 1.

of divergent interpretations of the text. When there are disagreements about what God is saying in the Bible—and the stakes
could not be higher—who has the authority to determine the
correct view? If there is no authority on earth that can adjudicate,
then the disputants remain in disagreement.

History has repeatedly shown that when no such authority is
recognized, the Christian community can fracture into separate
churches. This is a scandal and an embarrassment, for Christ's
prayer for his followers was "that they may all be one; even as
you, Father, are in me, and I in you, that they also may be in us,
so that the world may believe that you have sent me."[2]

The Catholic Church accepts that Christ instituted the apostles and their successors as this authority, to teach and preach in
his name. We can hear this from Jesus in the gospels of John and
Matthew, for example:

> Jesus said to them again, "Peace be with you. As the
> Father has sent me, even so I send you." And when he
> had said this, he breathed on them, and said to them,
> "Receive the Holy Spirit. If you forgive the sins of any,
> they are forgiven; if you retain the sins of any, they are
> retained."[3]

> Jesus came and said to them, "All authority in heaven and
> on earth has been given to me. Go therefore and make
> disciples of all nations, baptizing them in the name of the
> Father and of the Son and of the Holy Spirit, teaching
> them to observe all that I have commanded you."[4]

---

2. Jn. 17:21.
3. Jn. 20:21–23.
4. Matt. 28:18–20.

And so the Second Vatican Council affirms that "the task of authentically interpreting the word of God, whether written or handed on, has been entrusted exclusively to the living teaching office of the Church, whose authority is exercised in the name of Jesus Christ."[5]

Some Christians are uncomfortable with the idea of the Church's mediation here because it seems to cut off the believer's direct contact with God—and who could possibly serve as an adequate substitute for God? These Christians would maintain that when they read the Bible faithfully, the Holy Spirit will reveal to them what the Scriptures mean, and so there is no need for any authority other than the Holy Spirit.

Now, it is true that no one needs anything in addition to God the Holy Spirit, but the Holy Spirit Himself has willed to give the community of believers a charism that he gives to no single individual, and he has willed to work through the Church's ministers as his own instruments. Since the Bible itself is a mediated and inspired message of God, it is not unreasonable that its interpretation would also be mediated but in some way inspired.

Here it is important to note that it is not the Church instead of God who interprets Scripture. It is God who guides, speaks, and interprets *through* the Church. Analogies for this abound. For instance, could we not just go straight to God without going to the Bible, which was produced by humans (at least as instruments)? Of course we could, but we have to go to the Bible because God willed to reveal himself through this humanly written word. Could God not create and nurture new human life without human parents? Of course he could, but God has willed to use human parents as instruments to accomplish this task. Could God not speak directly to everyone at all times without

---

5.   *Dei Verbum*, Chapter 2, §10.

apostles, prophets, teachers, patriarchs, or miracle workers? Of course he could, but God has willed to use human instruments throughout salvation history, such as Moses, who spoke to Pharaoh in God's name and led the chosen people out of Egypt by God's power. In the same way, God has willed that one, holy, catholic, and apostolic Church interpret the sacred Scriptures in his name.

But is there not a great risk that corrupt ministers might discredit the Church, whereby the very authority of Scripture and of Christ Himself would be brought into question? Yes, indeed. But God takes such risks over and over again. Jesus did not change his mind about sending the apostles forth to teach and preach in his name, even though one of the twelve had betrayed Him, and the rest—save one—had abandoned him in his hour of greatest need. God takes the same risks with (potentially bad) parents, who have a decisive role in nurturing their child's life and faith. Astonishingly, in his wisdom and love, God calls us sinners to be servants consecrated to him and to share in his own ministry.

We should also note that what Catholics claim for the pope and bishops is less than what some of our separated brothers in Christ claim for each Christian—namely an infallible authoritative reading of Scripture under the guidance of the Holy Spirit, anywhere at any time. In the Catholic understanding, while the Holy Spirit regularly leads individual believers to a correct grasp of the Scriptures, he gives an infallible authoritative interpretation only to the pope and bishops who have come together in a council to determine, after prayerful discernment, what must be affirmed and what must be denied, as they read the Bible according to the sacred Tradition of the Church, that is, according to what has been believed and handed on from the beginning.

But where does this sacred Tradition come from? The New Testament comes out of and is predated by the community of believers, who have a living tradition which Acts 9:2 refers to as the "Way." The new Christians "devoted themselves to the apostles' teaching and fellowship, to the breaking of bread and the prayers."[6] Let us recall that the Church flourished with a living tradition after the Lord's Resurrection for about twenty years before the earliest text of the New Testament was written and for likely seventy years before the latest book of the Bible was completed. Also, for forty days after his resurrection, i.e., until he ascended into heaven, Jesus taught the apostles, "speaking of the kingdom of God."[7] So, this tradition comes from the Lord Jesus and is developed and strengthened by the Holy Spirit. St. Paul instructs the Thessalonians to "hold to the traditions" that he has taught them[8] and commends the Corinthians for doing just that.[9]

A Catholic's confidence in the Church's mediatory role is based on the fact that the Holy Spirit did not only work through the Church when the Bible was being written and compiled, but he works in the Church no less right now!

John Baptist Ku, O.P.

---

6. Acts 2:42.
7. Acts 1:3.
8. See 2 Thess. 2:15.
9. See 1 Cor. 11:2.

CHAPTER 14

Reading the Bible and the Senses of Sacred Scripture

Having established the authority of the Bible, its mediated character, and its authentic interpreter in the previous chapter, I will now examine the *senses of Scripture*, namely, the different manners in which one can read a particular passage in the Bible. Acknowledging that there may be numerous legitimate meanings of any given scriptural text will allow us to establish that we can read the creation narratives in different legitimate ways.

Following a long tradition, St. Thomas Aquinas recognizes a literal and a spiritual sense of Scripture.[1] These are often counted as "the four senses" of Scripture, i.e., the literal sense and three spiritual senses. But there are four possible divisions of the literal sense. Thus, by the time of Aquinas, the Catholic tradition counted seven senses.

St. Thomas follows St. Augustine in maintaining that "all the senses are founded on one sense, namely, the literal, from which alone can any argument be drawn, and not from those that are predicated according to the allegorical sense."[2] Thus the literal sense is primary and is the only one suitable to making properly theological arguments, e.g., about the Trinity, the Incarnation, or creation.

For Aquinas, the literal sense of any passage is what the author intends the words to mean. If the author intends to use a symbol in figurative speech, as in a metaphor or an allegory, then that figurative language is the literal meaning. For instance, if the author of the first creation account in Genesis intends the six days as a figure of speech in order to teach that God (and not a demon) created the material world, then the literal sense is not that the world was created in six days, but that God created the world. However, if the author intended the six days as six twenty-four hour periods, then such a timeframe would belong to the literal sense of the passage.

In order to understand the literal sense better, let us now distinguish it from the spiritual senses and review the four divisions it comprises. The three spiritual senses are the allegorical, the moral (also known as the tropological), and the anagogical.

In allegorical interpretation, which enjoyed immense popularity among patristic exegetes in the earliest centuries of Christianity, "things of the Old Law signify things of the New Law."[3] For instance: Isaac, who was to be sacrificed by his father; Joseph, who was sold into slavery by his brothers and fed his people in the time of famine; and Jonah, who spent three days in the belly of the whale, all signify Christ. As another illustration, the Passover, wherein an unblemished male lamb is sacrificed so that one might consume the lamb's flesh and place the lamb's blood on one's door lintels in order to find protection from the angel of death, signifies the Last Supper and the Crucifixion. And as a final example, the Exodus, where God's people miraculously marched through the

---

1. See *Catechism of the Catholic Church*, §115. The spiritual sense also goes by the name "mystical" sense (St. Thomas Aquinas, *Commentary on Galatians*, c. 4, lect. 7).

2. *ST* I, q. 1, a. 10, ad 1.

3. *ST* I, q. 1, a. 10.

sea into freedom, signifies the salvation worked through baptism. The allegorical sense is a true sense of Scripture, and allegorical interpretation has often been brilliant. But, as Aquinas explains, only the literal sense of the text can be the basis of theological argument. The language of the allegorical sense cannot be brought into conformity with the demands of syllogistic reasoning, since the meanings of its terms are not clearly fixed.

According to the moral sense, "things done in Christ or signifying Christ [in the sacred Scriptures] signify things we ought to do."[4] Why does Aquinas not simply say that things that Christ did signify what we should do? Because these exemplary actions of Christ are recommended to us under the literal sense, which of course can include moral exhortations or imperatives. Let us recall that if it was *the author's intention* for the reader to imitate Christ's action, such as carrying his cross or welcoming children, then taking Christ's action as signifying what we should do would fall under the literal and not the spiritual sense. The spiritual sense is naturally more symbolic and imaginative, as evidenced by two examples that Aquinas himself mentions. First, in the *Catena Aurea*,[5] Aquinas reports Rabanus' interpretation of a passage according to the moral sense: "The Cross signifies cheerfulness of action in its width, because sadness makes narrow. For the width of the Cross is in the transverse beam, where the hands are pierced, and through hands we understand acts." Second, in his commentary on Galatians,[6] Aquinas clarifies that "if it should be said that 'let there be light' means that it is through Christ that we should be illuminated in understanding and inflamed in love, it pertains to the moral sense."

---

4.   *ST* I, q. 1, a. 10.
5.   Chapter 27, lect. 7.
6.   Chapter 4, lect. 7.

In the anagogical sense, things in "the New and the Old Testaments at the same time signify the Church triumphant,"[7] or pertain "to eternal glory."[8] For instance, "if it should be said that 'let there be light' means that through Christ we should be led to glory, it pertains to the anagogical sense."

So, these three senses: the allegorical, the moral, and the anagogical, are examples of the spiritual sense of Scripture, where the interpretation goes beyond the literal sense, i.e., what the author intended the words to mean. Even so, as St. Thomas clarifies, all senses of Scripture must be based on the literal sense, and theological arguments can only be drawn from the literal sense. Aquinas identifies four possible divisions of literal sense as: the historical, the etiological, the analogical, and the parabolical (also known as the metaphorical) senses.

In the historical sense, something is simply reported as having happened. So, Joseph was sold into slavery, and the Jews were enslaved by the pharaoh who knew not Joseph. Those are historical facts.

The etiological sense "assigns the cause of what is said" in the Bible, such as when Jesus gave the reason why Moses allowed wives to be divorced: the Lord revealed that it was on account of the hardness of men's hearts.[9]

In the analogical sense, "the truth of one Scripture is shown not to contradict the truth of another."[10] Here it is recognized that since God is the primary author of the Scriptures, despite the number and diversity of the books that constitute the Bible, there is one consistent message that is being communicated. So,

---

7.   *Quodlibet* VII, q. 6, a. 2.

8.   *ST* I, q. 1, a. 10.

9.   Matt 19:8. See *ST* I, q. 1, a. 10, ad 2.

10.   *ST* I, q. 1, a. 10, ad 2.

when Paul talks about being saved by faith and not by works, and James asserts that faith without works is dead,[11] we recognize that the literal meaning of "works" can be different in different contexts: Paul is discussing works of the Law, while James has works of charity in mind. Also, as we will see in the next chapter, if two creation accounts contradict each other in the reporting on the order of events, then the details of this order of events cannot be the literal sense of these texts—at least not of them both.

The parabolical sense refers to the author's use of symbols. That is, the literal sense includes symbolic and metaphorical usage of words. For instance, when Scripture speaks of "God's arm,"[12] the author means to signify *God's power*, and does not believe that he who is pure spirit[13] has skin, bones, and a really big elbow.

In conclusion, of the two senses of Scripture, literal and spiritual—which can be subdivided into seven senses in all, three spiritual and four literal—only the literal can be the basis of a theological argument. That means that arguments about creation bearing on evolution must appeal to the *literal sense* of the biblical creation accounts, which can include symbolic language and figures of speech if it was the author's intention to use such language in the passage.

John Baptist Ku, O.P.

---

11. Rom. 3:28 and Jam. 2:17.
12. E.g., Dt. 4:34, Is. 40:10, Jer. 32:17, and Acts 13:17.
13. Jn. 4:24.

CHAPTER 15

Interpreting the Creation Narratives in the Bible

The six-day creation account in the first chapter of Genesis is often brought to bear on the theological debate over evolution to the exclusion of other creation narratives in the Bible. This is not surprising because this account is related to us in the very first words of the Bible, and, at face value, the whole point of this narrative seems to be to spell out God's creative activity in a detailed record of six days: on the first day God created light; on the second day, the heavens; on the third day, the land and plants; on the fourth day, the sun, the moon, and the stars; on the fifth day, birds and fish; and on the sixth day, land animals, and a man and a woman. For many, it is difficult to reconcile this picture of creation with the theory of evolution.

To interpret the Genesis 1 pericope correctly, we must read it in the context of the rest of the Bible, which contains other creation accounts that are often overlooked.

For example, there is a second creation narrative in Genesis, which appears in chapter 2. This account presents a different order of creation: Eve is made from Adam's rib only after the animals have been created and are found to be unworthy partners for the man. In this reportage, God made the heavens and the earth, and then the first man, Adam. Whether this all occurred on the same day or on different days is not specified. God then planted a garden and placed the man in it. Next, God made the animals, which Adam named as each was brought to him. And finally, Eve was created.

Now, if the literal sense of these passages in Genesis 1 and 2 were to present the historical order of the production of creatures, then they would be in contradiction; and thus Scripture would not be inerrant. For in the first account, the man and woman were created together after the animals while in the second account, the man was created before the animals, but the woman was made last of all creatures.

However, if the literal sense is rather to affirm that God alone creates something where before there was nothing, that there is a divinely intended order among all corporeal creatures with man at the top, that God made Adam and Eve for each other—and other such theological themes—then there is no contradiction since the author never intended to provide an astronomical or zoological record of an event at which he was not present. Let us recall from the previous chapter that the literal sense includes symbols and figures of speech when it is the author's intention to use them.

Joseph Cardinal Ratzinger, now Pope Emeritus Benedict XVI, offers a teaching on this question in a homily published in his insightful book *In the Beginning*:

> The Scripture would not wish to inform us about how the different species of plant life gradually appeared or how the sun and the moon and the stars were established. Its purpose ultimately would be to say one thing: *God* created the world. The world is not, as people used the think then, a chaos of mutually opposed forces; nor is

it the dwelling of demonic powers from which human beings must protect themselves.[1]

What about other creation accounts? We find important teachings on creation in the historical books (2 Maccabees), in the wisdom literature (i.e., Job, Psalms, Proverbs, Wisdom, and Sirach), in the prophets (i.e., Isaiah), and in the New Testament (i.e., John, Romans, Colossians, and 2 Peter).[2] These are not texts whose sole purpose is to describe the beginning of the world as we find in Genesis, but they constitute definitive revelation about the divine act of creation.

We will now closely examine a passage from Proverbs with the help of selections from John and Colossians, and then simply draw together some other relevant texts to display a sample of Scripture's treatment of creation. By considering these sources that are often overlooked in the debate over evolution, we will benefit from a fuller perspective.

In Proverbs 3:19–20 we are treated to a more theological explanation of creation, namely, that God created everything *by his wisdom*. The preceding verses (vv. 13–17) situate this passage in the context of the praise of wisdom: that is, the man who finds wisdom is "happy" for "she is more precious than jewels, and nothing you desire can compare with her." Were that not enough, "long life is in her right hand; in her left hand are riches and honor" and "her ways are ways of pleasantness, and all her paths are peace." The next verse (v. 18) makes an allusion to the

1. Pope Benedict XVI, *In the Beginning* (Grand Rapids: Eerdmans, 1995), Homily 1, p. 5.
2. 2 Mac. 7:28; Job 9:8–9; 38:24–25; Ps. 8:3, 6–8; Ps 19:1, 4; Ps. 33:6–7, 9; Ps. 65:6–8, 104:2–46, 139:13–14, 147:4–18, 148:5–10; Prov. 8:24–30; Wis. 1:14, 2:23, 11:17; Sir. 16:26–27, 17:1; Is. 42:5, 44:24, 45:7, 45:12, 45:18, 48:13; Rom. 4:17.

creation account in Genesis 2, calling wisdom "a *tree of life* to those who lay hold of her."

Then in verses 19–20, we come to the profound theological assertion that "YHWH by wisdom founded the earth; by understanding he established the heavens; by his knowledge the deeps broke forth, and the clouds drop down the dew." This is clarified in John 1:1–3, which restates the classic creation account of Genesis 1 in terms of the uncreated Word of the Father, that is, the Father's Wisdom or Knowledge, through whom all things were made: "In the beginning was the Word, and the Word was with God, and the Word was God. He was in the beginning with God; all things were made through him, and without him was not anything made that was made." St. Paul advances the same teaching in the early part of his letter to the Colossians: "For in him [the most beloved Son, the image of the invisible Father] all things were created: things in heaven and on earth, visible and invisible, whether thrones or powers or rulers or authorities; all things have been created through him and for him. He is before all things, and in him all things hold together."[3] But what does this mean?

St. Thomas Aquinas borrows an image from Aristotle in order to explain this doctrine of acting through wisdom. The carpenter who makes a bench does so not only with wood and saw and hammer. He also must have *the idea of the bench* he wants to make. So, the bench is made *through his knowledge of the bench*, that is, *through his wisdom*. If he is not wise about benches—for instance, if has no idea what it should look like—then he cannot make it even if he has the best wood, saw, and hammer in the world. It is correct in the strictest and most theologically rigorous sense to say that the Father *creates all things through his Word*, the beloved Son, who proceeds in the divine intellect as begotten

---

3. Col. 1:16–17. See also 2 Pet. 3:5.

wisdom. The Father in knowing Himself (the divine essence) produces a concept of Himself, which is a perfect reflection of Himself, namely the Word, who is his Son. And since the act of creation proceeds from God's knowledge of what he wishes to create, *all things were made through the Word.*

So, while the central point of Proverbs 3:13–20, John 1:1–3, and Colossians 1:16–17 may not be to explain the production of creatures, these passages do articulate correctly and profoundly how all things came to be.

Two other passages are theologically important for their clear assertion that God created "out of nothing" (*ex nihilo*), that is, not from preexisting chaos or unformed matter but where before there was simply nothing. In 2 Maccabees 7:28, we are instructed to "look at the heaven and the earth and see everything that is in them, and recognize that God did not make them out of things that existed." And Paul explains to the Romans that God "gives life to the dead and calls into existence the things that do not exist."[4]

Six other groups of passages merit mention simply because they speak of creation at some length and should thus not be passed over. First, in Proverbs, wisdom is again connected to the act of creation, i.e., wisdom preexisted creation and was involved in its production: "When there were no depths I was brought forth, when there were no springs abounding with water. Before the mountains had been shaped, before the hills, I was brought forth; before he had made the earth with its fields, or the first of the dust of the world. When he established the heavens, I was there, when he drew a circle on the face of the deep, when he made firm the skies above, when he established the fountains of the deep, when he assigned to the sea its limit, so that the

---

4. Rom. 4:17.

waters might not transgress his command, when he marked out the foundations of the earth, then I was beside him, like a master workman; and I was daily his delight, rejoicing before him always."[5]

Second, the Psalms too speak of God's creation through his word: he establishes the world, overcomes chaos, and provides for all his creatures. For example, Ps. 33:6–7, 9, teaches that YHWH created everything by his word and imposed boundaries on the sea: "By the word of YHWH the heavens were made, and all their host by the breath of his mouth. He gathered the waters of the sea as in a bottle; he put the deeps in storehouses. For he spoke, and it came to be; he commanded, and it stood forth." God's mastery over the chaotic waters, which will demonstrate Christ's divinity in the gospels, is confirmed in Ps. 65:6–8: "O God...who formed the mountains by your power, having armed yourself with strength, who stilled the roaring of the seas, the roaring of their waves, and the turmoil of the nations. The whole earth is filled with awe at your wonders; where morning dawns, where evening fades, you call forth songs of joy. You care for the land and water it; you enrich it abundantly." And a short sample from Psalm 104's thirty-one-verse section about God's creation of and providing for his creatures suffices to manifest the character of this theological poem: "The trees of the Lord are well watered, the cedars of Lebanon that he planted. There the birds make their nests; the stork has its home in the junipers. The high mountains belong to the wild goats; the crags are a refuge for the hyrax. He made the moon to mark the seasons, and the sun knows when to go down."[6]

---

5.   Prov. 8:24–30.
6.   Verses 16–19. See also Ps. 8:3, 6–8; Ps. 19:1, 4; Ps. 104:2–32; Ps. 139:13–14; Ps. 147:4–18; Ps. 148:5–10.

Third, Sirach 16:26–27 emphasizes the order in creation: "The works of the Lord have existed from the beginning by his creation, and when he made them, he determined their divisions. He arranged his works in an eternal order, and their dominion for all generations." And Sirach 17:1 repeats the theme of the second creation account in Genesis, i.e., that "the Lord created man out of earth, and turned him back to it again."

Fourth, Wisdom 1:14 highlights God's power and the goodness of creation, themes already present in Genesis: "For he created all things that they might exist, and the generative forces of the world are wholesome, and there is no destructive poison in them; and the dominion of Hades is not on earth."[7] And Wisdom 2:23, already hinting at the resurrection, adds the character of incorruptibility to man's being in image of God: "for God created man for incorruption, and made him in the image of his own eternity."

Fifth, in Job 9:8–9, God again stamps out chaos: "who alone stretched out the heavens, and trampled the waves of the sea; who made the Bear and Orion, the Pleiades and the chambers of the south." And throughout chapter 38, God's transcendent creative power is affirmed in the hard questions posed to Job, who pretended to understand how things should have gone: "What is the way to the place where the light is distributed, or where the east wind is scattered upon the earth? Who has cleft a channel for the torrents of rain, and a way for the thunderbolt?"[8]

Sixth, and finally, Isaiah 45:18 emphasizes that it is God alone who has the power to create, and here too, he overcomes chaos: "For thus says YHWH, who created the heavens (he is God!), who formed the earth and made it (he established it; he

---

7. See also Wis. 11:17.
8. Job 38:24–25.

did not create it a chaos, he formed it to be inhabited!): 'I am YHWH, and there is no other.'"⁹

Joseph Cardinal Ratzinger comments on "the fact that the classic creation account [of Genesis 1] is not the only creation text of sacred Scripture," noting that "in its confrontation with Hellenistic civilization, Wisdom literature reworks the theme without sticking to the old images such as the seven days." "Thus," he explains, "we can see how the Bible itself constantly readapts its images to a continually developing way of thinking, how it changes time and again in order to bear witness, time and again, to the one thing that has come to it, in truth, from God's Word, which is the message of his creating act. In the Bible itself the images are free and they correct themselves ongoingly." "In this way," he continues, "they show, by means of a gradual and interactive process, that they are only images, which reveal something deeper and greater."¹⁰

In conclusion, it is important to note that there is not just one creation account in the Bible that describes a six-day process. There is a second narrative in Genesis, and there is a more theological explanation in the historical books, the wisdom literature, the prophets, and the New Testament. Conflicting assertions between the first and second creation stories in Genesis with respect to the order of events manifest that the literal sense of those details cannot be a scientific reportage—if the Scriptures are the inerrant word of God.

John Baptist Ku, O.P.

---

9. Is. 16:26–27.
10. *In the Beginning,* Homily 1, pp. 14–15.

❧ CHAPTER 16

Interpreting Genesis One with the Fathers of the Church

In this chapter, we will consider the interpretation of the creation account in Genesis 1 offered by various Church Fathers. The gain of this undertaking will be the observation that there are different opinions among the greatest saints and scholars in the Church. By taking note of this, we can avoid a myopic reading of Genesis and thus steer clear of an uninformed fear-driven reaction to atheistic materialistic assertions about the origin of the universe.

We will briefly review the positions of twelve Church doctors and then consider St. Augustine of Hippo's understanding at greater length. Augustine is a good choice, not only because he is one of the most influential figures in the Church, but also because he is exemplary in explicitly joining a defense of the truth of Scripture with a caution about drawing overly detailed conclusions about this mysterious event that predates human existence.

A distinction that must be noted preliminarily is the difference between the act of creation properly speaking, which is to produce something where before there was simply nothing, and the act of making something more interesting out of basic elements that already exist.[1] With the benefit of ancient Greek philosophy, some of the earliest Fathers already articulate this difference, observing that the act of creation properly speaking must be instantaneous, leaving the question of the six days to be a matter of interpreting the formation of the basic elements created out of nothing.

Some Fathers of the Church suggest a reading of the six days as twenty-four hour periods. For examples of this, we can point to St. Victorinus who accepts that "God produced the entire mass for the adornment of his majesty in six days," clarifying a few lines later that "in the beginning God made the light, and divided it in the exact measure of twelve hours by day and by night."[2] Also, Lactantius in his *Divine Institutes* seems to accept six ordinary days, which he correlates to six ages of a thousand years each, after which the world will end.[3]

Turning to slightly later, more significant authorities, Ss. Basil the Great, Ambrose of Milan, and John Damascene explicitly refer to twenty-four hour periods. In speaking of the "first day," Basil explains that "it is as though [Scripture] said: twenty-four hours measure the space of a day, or that, in reality a day is the time that the heavens starting from one point take to return there."[4] And Ambrose, also commenting on the first day, asserts that "Scripture established a law that twenty-four hours, including both day and night, should be given the name of day only,

---

1. See, for instance, John Damascene's *De Fide Orthodoxa*, 2.5: "Our God himself, whom we glorify as Three in One, *created the heaven and the earth and all that they contain*, and brought all things out of nothing into being: some he made out of no preexisting basis of matter, such as heaven, earth, air, fire, water: and the rest out of these elements that he had created, such as living creatures, plants, seeds. For these are made up of earth, and water, and air, and fire, at the bidding of the Creator."

2. *On the Creation of the World* (written c. AD 280).

3. *Divine Institutes*, Book 7, Chapter 14 (written c. AD 307).

4. *Hexaemeron*, Homily 2, No. 8 (written c. AD 370).

as if one were to say the length of one day is twenty-four hours in extent."[5] John Damascene describes a day's length in terrific detail, comparing the solar solstice to the solar equinox, and the period of the moon to the sun, when he ponders the sun's being created on the fourth day.[6] However, to recall our previously noted distinction, Basil, Ambrose, and Damascene also teach that the universe was created instantaneously or before time.[7]

We also find ancient Doctors of the Church who do not suppose that a day in the Lord's time is twenty-four hours. Ss. Justin Martyr and Irenaeus of Lyons quote the line "The day of the Lord is a thousand years" from Psalm 90:4 in connection with Adam's dying on "the same day" that he ate the apple.[8] And St. Cyprian writes that "The first seven days in the divine arrangement contain seven thousand years."[9]

Along the same lines, but with deeper metaphysical considerations, St. Clement of Alexandria and Origen both recall Genesis 2:4: "In *the day* that the LORD God made the earth and the heavens," as evidence that the "six days" are to be taken figuratively.[10] In his *Miscellanies*, Clement notes that the creation could not have taken place in time because time itself was created.[11] So, new things could be "generated" over a span of days, but creation itself did not transpire over a period of time but is rather the source of time.

---

5.   *Hexaemeron*, Chapter 10, No. 37 (written in AD 393).

6.   *De Fide Orthodoxa*, Book 2, Chapter 7 (written c. AD 710).

7.   St. Basil, *Hexaemeron*, Homily 1, No. 6; St. Ambrose, *Hexaemeron*, Chapter 10, No. 37; St. John Damascene, *De Fide Orthodoxa*, Book 2, Chapter 1.

8.   Justin Martyr, *Dialogue with Trypho the Jew* 81 (written c. AD 155); Irenaeus of Lyons, *Against Heresies* 5.23.2 (written c. AD 189).

9.   *Treatises* 11.11 (written c. AD 257).

10.   St. Clement of Alexandria, *Miscellanies* 6.16 (written c. AD 208); Origen, *Against Celsus* 6.51–61 (written c. AD 248).

11.   *Miscellanies* 6.16.

Origen argues similarly that "there was not yet time before the world existed,"[12] and that the first days cannot be taken literally because you cannot have a day without a sun, a moon, and a sky.[13] So, early in the third century Clement and Origen have already articulated the central difficulties in taking six ordinary days as the literal sense of Genesis 1.

St. Augustine does not interpret the six days of creation to be six periods of twenty-four hours. He treats this theme in a few different works and is consistent on this point. In his *Commentary on the Book of Genesis*, Augustine makes some important distinctions about how Scripture should be read and then directly addresses the question of creation in Genesis 1:

> In all the sacred books [of Scripture], we should consider
> the eternal truths that are taught, the facts that are narrat-
> ed, the future events that are predicted, and the precepts
> or counsels that are given. In the case of a narrative of
> events, the question arises as to whether everything must
> be taken according to the figurative sense only, or wheth-
> er it must be expounded and defended also as a faithful
> record of what happened. No Christian will dare say that
> the narrative must not be taken in a figurative sense. For
> St. Paul says: "Now all these things that happened to them
> were symbolic." And he explains the statement in Genesis,
> "And they shall be two in one flesh," as a great mystery in
> reference to Christ and to the Church. If, then, Scripture
> is to be explained under both aspects, what meaning oth-
> er than the allegorical have the words: "In the beginning

---

12. *Homilies on Genesis* (written c. AD 234).
13. *Fundamental Doctrines* 4.1.16 (written c. AD 225); *On First Principles* 4.3.1 (written c. 215).

God created heaven and earth?" Were heaven and earth made in the beginning of time, or first of all in creation, or in the Beginning who is the Word, the only-begotten Son of God? And how can it be demonstrated that God, without any change in Himself, produces effects subject to change and measured by time? And what is meant by the phrase "heaven and earth"? Was this expression used to indicate spiritual and corporeal creatures?[14]

Augustine goes on to raise a host of possible interpretations of "heaven," "the earth," "darkness," "the abyss," and "let there be light"; and he will conclude that "heaven and earth" refer to formless matter.[15] He explains that in a narration, you must give one thing before the other, but that doesn't mean that there is a difference in time. So, the first day and the second day are not different times but different orders. He offers the example of speech: "But the speaker does not first utter a formless sound of his voice and later gather it together and shape it into words. Similarly, God the Creator did not first make unformed matter and later, as if after further reflection, form it according to the series of works he produced. He created formed matter." Augustine clarifies that the material itself of a thing does in a certain sense precede the thing, as clay in a certain sense precedes a clay pot; but whatever shape the clay has at any time is simultaneous with its being clay.[16]

In a manner reminiscent of Origen's argument, Augustine doubts the counting of six ordinary days, pointing out that the sun would never set on God in his creation—for where would it

---

14. *Commentary on the Book of Genesis*, Book 1, Chapter 1, Nos. 1–2 (completed in AD 415).

15. *On Genesis*, Book 1, Chapter 14.

16. *On Genesis*, Book 1, Chapter 15, No. 29.

go, to another universe?[17] And not unlike Clement, Augustine insists that creation had to be instantaneous:

> No one certainly would be so foolish as to think that, because God is great beyond all beings, even a very few syllables uttered by his mouth could have extended over the course of a whole day. Furthermore, it was by his coeternal Word, that is, by the interior and eternal forms of unchangeable Wisdom, not by the material sound of a voice, that "God called the light Day and the darkness Night." And further questions arise. If he called them with words such as we use, what language did he speak? And what was the need of fleeting sounds where there was no bodily sense of hearing? These difficulties are insurmountable in such a supposition.[18]

Helpful to our overall consideration, Augustine warns against pretending to have one single right interpretation of these difficult passages:

> In matters that are obscure and far beyond our vision, even in such as we may find treated in Holy Scripture, different interpretations are sometimes possible without prejudice to the faith we have received. In such a case, we should not rush in headlong and so firmly take our stand on one side that, if further progress in the search of truth justly undermines this position, we too fall with it. That would be to battle not for the teaching of Holy Scripture but for our own, wishing its teaching to conform to ours,

---

17. *On Genesis*, Book 1, Chapter 10, No. 21.
18. *On Genesis*, Book 1, Chapter 10, No. 20.

whereas we ought to wish ours to conform to that of sacred Scripture.[19]

Augustine presciently adds that we only damage Scripture's credibility—especially in the minds of unbelievers who are educated in science—if we draw wrong conclusions about science from the Bible: "Now, it is a disgraceful and dangerous thing for an infidel to hear a Christian, presumably giving the meaning of Holy Scripture, talking nonsense on these topics."[20]

Finally, let it not be thought that Augustine thinks that Christians should cower before and defer to irreligious experts in science: "With a sigh, [such weak believers] esteem these teachers as superior to themselves, looking upon them as great men; and they return with disdain to the books which were written for the good of their souls; and, although they ought to drink from these books with relish, they can scarcely bear to take them up. Turning away in disgust from the unattractive wheat field, they long for the blossoms on the thorn."[21]

To sum up then, it is clear that different Fathers of the Church interpreted the first chapter of Genesis in diverse ways. They certainly did not all interpret the literal sense of the six days to be six twenty-four hour periods, as Augustine, one of the greatest theologians of the Church, shows. And in the influential Fathers who do so interpret the six days, we find the distinction between creation, which is instantaneous and before time, and the subsequent development of that creation over six days.

John Baptist Ku, O.P.

---

19. *On Genesis*, Book 1, Chapter 18, No. 37.
20. *On Genesis*, Book 1, Chapter 19, No. 39.
21. *On Genesis*, Book 1, Chapter 20, No. 40.

❧ CHAPTER 17

Interpreting Genesis One with
St. Thomas Aquinas

In this chapter, we will review St. Thomas Aquinas' interpretation of the account of creation that appears in the first chapter of the Book of Genesis. Like Augustine, Aquinas was one of the most influential theologians in the Church; and with the advantage of his predecessors' excellent work, he was able to produce a coherent theological synthesis that stands even today as the authority of first recourse for difficult theological questions.

We will see in this chapter that Aquinas has carefully studied older opinions about Genesis 1, but does not choose one over another. Instead he shows where they agree or disagree, what must be ruled out, and why we can accept differences of opinion where we in fact can. In his own reading of Genesis 1, Aquinas distinguishes three phases within the six days of creation, but he does not discuss whether the days are twenty-four hour periods or are rather symbols of different orders of creatures.

Like the Fathers of the Church, Aquinas observes the important metaphysical distinction between creation and change. Creation is the act of making things exist where before nothing existed. In contrast, forming a new interesting thing out of preexisting basic elements is an impressive *change*, but it is not creation properly speaking. Coincidentally for evolution, the word Aquinas uses for "change" is *mutatio*: he repeatedly reminds his readers that *creatio non est mutatio*—that is, creation is not change, alteration, development, or mutation.[1]

Aquinas clarifies that the act of creation requires omnipotence, so there are no intermediate actors in creation.[2] Also, as St. Thomas points out, the act of creation itself is indivisible, i.e., it does not take any time. God instantaneously and effortlessly wills the universe to exist—and so it did and does.[3]

Regarding the interpretation of the first chapter of Genesis, Aquinas manifests a detailed knowledge of the Fathers' diverse opinions (e.g. whether the firmament is the heavens, whether the empyrean heaven is the starry heavens; how water, land, air, and vapors develop, etc.[4]), and he appreciates the insights of them all. In his earliest theological synthesis, the *Commentary on the Sentences of Peter Lombard,* Thomas notes that the view that the world developed over six ordinary days "is the more common position and seems more consonant with the letter [of the text] on a superficial level."[5] But he judges that St. Augustine's understanding of the six days as signifying different orders of creatures and not different periods in time "is more rational and better defends sacred Scripture against the mockery of unbelievers." Aquinas says that he "likes this [latter] opinion more" but that "nevertheless all of the arguments can be answered in holding either opinion."[6]

---

1. II *Sent.*, d. 1, q. 1, a. 2, ad 2; *SCG* II, ch. 17–18, 20; *ST* I, q. 45, a. 2, ad 2; *ST* I, q. 46, a. 2, ad 1; *DP*, q. 3, a. 2, s.c.

2. I *Sent.*, d. 21, q. un., a. 1, qla. 1; I *Sent.*, d. 43, q. 1, a. 2, s.c. 1; II *Sent.*, d. 1, q. 1, a. 4; *ST* I, q. 65, a. 3; *ST* III, q. 13, a. 2.

3. I *Sent.*, d. 8, q. 3, a. 2, corp.; I *Sent.*, d. 43, q. 1, a. 2, s.c. 1; II *Sent.*, d. 15, q. 3, a. 3, obj. and ad 4; *ST* I, q. 74, a. 1, ad 1; *DP*, q. 5, a. 1, s.c. 2.

4. For example, see his *De Potentia,* q. 4, a. 1, ad 15.

5. II *Sent.*, d. 12, q. un., a. 2, corp.

6. II *Sent.*, d. 12, q. un., a. 2, corp.

In his *Summa Theologiae*, a later work that represents his last word on the matter, when Aquinas entertains the question *whether the firmament was made on the second day*, he begins by repeating St. Augustine's teaching that two things are important regarding the interpretation of such passages in Scripture:

> First, that the truth of Scripture be held without wavering. Second, that since sacred Scripture can be explained in many ways, one should adhere to no explanation so precipitously that he would [still] presume to assert this understanding of Scripture [even if] it were [later] agreed, because of a certain argument, that this position is wrong—lest Scripture be mocked by unbelievers because of this, and the way of believing be blocked for them.[7]

After reviewing various positions that take "day" as a twenty-four-hour period, Aquinas then observes:

> But if by these "days" the succession of time is not indicated but only an order of nature, as Augustine would have it, [still] nothing would prohibit our saying in agreement with any of these opinions [above] that the formation of the substance of the firmament pertains to "the second day."[8]

So, St. Thomas advises us not to overcommit ourselves to non-necessary opinions that might later be definitively refuted, and he looks for common ground within opposed acceptable views.

---

7. *ST* I, q. 68, a. 1.
8. Ibid.

Along the same lines, later in this same section on the work of the six days, Aquinas devotes an article to Augustine's claim that the six days are really one "day," or event. Conceding that Augustine's opinion is quite different from that of others on four points, Aquinas draws out the consistency between these contrasting approaches by showing how the divergent conclusions flow from different premises while the understanding of the *manner of creatures' production* is not so different. And whose side does Aquinas take? In the last sentence of his argument Aquinas declares that "in order to judge this from an unbiased position, we must reply to the arguments of both sides."[9]

As we would expect, St. Thomas does not think that all theological opinions are acceptable, or that it is simply a matter of personal preference. Returning to his *Commentary on the Sentences*, we find his sharp distinction concerning how conflicting interpretations of Scripture should be treated:

> Those things that pertain to the faith are distinguished in two ways. For certain things are of themselves the substance of the faith, such as that God is three and one, and this kind of thing, in which *no one is permitted to opine otherwise....* But other things are only accidentally the substance of the faith, insofar namely as they are handed on in Scripture...such as many historical facts which can, without danger, be unknown by those who are not obligated to know. And on this kind of facts even the Fathers held diverse opinions, explaining sacred Scripture diversely. Thus concerning the beginning of the world, there is something that pertains to the substance of the faith, namely, that the created world had a beginning,

9.   *ST* I, q. 74, a. 2.

and all the Fathers agree on this. But how it began and in what order it was made pertain to the faith only accidentally, insofar as these opinions are handed on in Scripture, whose truth the Fathers, holding diverse opinions, handed on by diverse explanations.[10]

Aquinas freely uses the language of Genesis, such as that the sun was created on *the fourth day*, but he leaves the term in its original poetic ambiguity. In his entire corpus, nowhere does he use the expression "twenty-four hour(s)," and as we have seen, he does not take a position on the question, because neither conclusion can be proven, and the point is not theologically decisive.

Even so, St. Thomas does comment on the Genesis 1 account. In his prologue to q. 65 in the first part of his *Summa Theologiae*, where he begins his treatment of the six days of creation, Aquinas distinguishes three phases:

[1] the work of creation, as given in the words, "In the beginning God created heaven and earth"; [2] the work of distinction as given in the words, "He divided the light from the darkness, and the waters that are above the firmament from the waters that are under the firmament"; and [3] the work of adornment, expressed thus, "Let there be lights in the firmament."

A few questions later in the *Summa Theologiae*, we find a succinct summary of Thomas' overall position on this issue, i.e., that the idea of creation over a period of time must be ruled out, that it is acceptable but not necessary to hold that there was special development over six days after creation, and consequently, that

---

10. II *Sent.*, d. 12, q. un., a. 2, corp.

it makes sense to speak of phases of creation, distinction, and adornment:

> According to Augustine, the work of creation pertains to the production of unformed matter and of unformed spiritual nature, both of which are outside of time, as he says in Book 12 of the *Confessions*. Therefore, the creation of either is placed before any "day." But according to some Fathers, it can be said that the work of distinction and adornment is applied according to some change in creatures, which is measured by time. But the work of creation consists in the divine action alone in the instant of producing the substance of things. And therefore, any work of distinction and adornment is said to be done "in a day," but creation is said to be done "in the beginning," which bespeaks something indivisible.[11]

Although Aquinas does not take St. Augustine's view to the exclusion of others on the question of what a "day" in Genesis 1 means, his understanding of the act of creation unfolding in two later phases of differentiation and perfection follows Augustine's lead. At times quoting Augustine explicitly, Aquinas speaks of seminal essences or principles given in creation that blossom into full form later. Obviously he is not thinking of Darwinian evolution, but his thought is not incompatible with what modern science appears to confirm. Here are a few passages to this effect:

> God is said to have stopped creating new creatures on the seventh day, because nothing was made afterwards that did not come first in some likeness according to genus or

---

11. *ST* I, q. 74, a. 1, ad 1.

species, at least in a seminal principle.... Therefore, I say that the future renewal of the world indeed came first in the works of the six days in a remote likeness.[12]

For thus we see that all things that were produced in the process of time through the work of Divine Providence, with creation operating under God, were produced in the first condition of things according to certain seminal patterns, as Augustine says in his *Commentary on the Book of Genesis*, such as trees, animals, and other things of this kind.[13]

On the day on which God created heaven and earth, he also created every plant of the field, not actually, but before it should spring up upon the earth, that is, potentially. It was not because of any lack in God's power, as if he needed time to work, that all things were not simultaneously [created and] distinct and adorned, but so that the order would be preserved in the arrangement of things.[14]

Ultimately then, Aquinas cautions against overcommitting ourselves to non-necessary and potentially vulnerable theological positions concerning the interpretation of Genesis 1. And he works to show the important harmony among different views held by various Fathers of the Church.

John Baptist Ku, O.P.

---

12. IV *Sent.*, d. 48, q. 2, a. 1, ad 3.
13. *ST* I, q. 62, a. 3.
14. *ST* I, q. 74, a. 2, ad 1.

❧ CHAPTER 18

Modern Biblical Exegesis of the Creation Accounts

This chapter will consider the contributions of modern biblical scholarship to the interpretation of the creation accounts in Scripture. Although modern exegesis differs significantly from a Thomistic approach, it should not be omitted in a consideration of how to understand the creation narratives, for it can be used in harmony with Aquinas' approach to sketch out a more complete account.

One of the characteristics of modern biblical exegesis is a strong interest in ferreting out original sources. For instance, can we discern the hands of different authors in the Pentateuch? Or do creation accounts in the Bible borrow from older pagan creation myths?

One might be concerned that, by establishing that the biblical creation accounts are influenced by older pagan stories, one undermines biblical inspiration. That is, if the Bible has pagan sources, then it would seem that the text is not original and revealed by God, but cut and pasted from older manmade myths.

However, divine inspiration should not be equated with discontinuous originality. Before God chose his people Israel, making a covenant with Abraham, they were naturally like their neighbors. In revealing himself to his people, God not only gave them an entirely new revelation but also corrected and purified prevailing notions about Himself and the universe that the people already had. Thus if Israel's inspired religious poetry retains vestiges of a polytheistic past, such as "God of gods"[1] we need not be unsettled.

Moreover, we should not be startled that pagan peoples could make some right judgments about God and the world in their religious poetry. After all, as St. Thomas Aquinas teaches, one can know by reason alone that God exists and is the cause of all creatures.

Furthermore, if the faith is true, then it has nothing to fear from any scientific discovery—be it biological, astronomical, archeological, paleological, or philological. For God is the one source of all truth, whether revealed or discovered by reason. Therefore, if similar stories or the same names that we read in the Scriptures are discovered in more ancient pagan literature, there is no cause for concern.

Admittedly, some modern scholarly approaches to the Bible adopt a hermeneutic of suspicion. That is, they presuppose that the believing author or ecclesial redactor plays with the facts in order to bend our minds to his disingenuous religious agenda. The task of the interpreter then is to break through the layer of propaganda by getting behind the text so as to extract the pure facts. Yet even where investigation is contaminated by this hostile skepticism, the correct results of historical research into ancient cultures should be accepted.

In this chapter, we will briefly review a number of themes present in biblical creation accounts that modern Scripture scholarship finds in older pagan sources, namely: a personal creator,

---

1. Dt. 10:17, Ps. 84:7, Ps. 136:2, Dan. 11:36.

the impersonal production of things, creation through the separation of elements, creating by speaking a word, the formation of man out of clay, man's being made in the image of the maker, and conflict with chaos and dragons or sea monsters. From the examples presented, the reader will be able to judge for himself how close the connections are between the Bible and the more ancient myths. In any case, the evidence should suffice to show that, while there was an original and unique revelation made to Israel, the creation accounts do not articulate this revelation in isolation from Israel's neighbors.

Scholars have established that thousands of years before the nation of Israel was constituted, the peoples in that region already had a personal creator god in their mythology. What is unique about the Old Testament is that God is categorically outside of creation; he is *uncreated*. Other creation accounts include gods within creation, even if they are the highest or most powerful in the universe.[2] This will be illustrated in the examples below. To situate the chronology, we should note here that the great Babylonian epics like the Gilgamesh story (c. 1200 B.C.) and the Enuma Elish (c. 1700 B.C.) are drawn from older individual Sumerian myths that were later integrated.[3] The oldest sources of the Pentateuch are believed to be from around 900 B.C.[4]

In ancient creation myths there is a transition from a description of the spontaneous production of things to the attribution of their production to a personal god. At an even later stage, praise is offered to the creator god. Modern biblical scholars observe vestiges of the more ancient origin stories in the Bible.

---

2. Claus Westermann, *Genesis: An Introduction*, p. 25. This essay will take Westermann to be representative of modern Scripture scholarship. He documents other authors abundantly, signaling where he agrees or disagrees with them.

3. Ibid., p. 23.

4. *The Oxford Dictionary of the Christian Church*, 3rd ed., s.v. "Pentateuch."

For instance, in Gen. 1:24: "And God said, 'Let the earth bring forth living creatures according to their kinds,'" one sees the language of the spontaneous production of things, now set in the context of God's commanding it. And Ps. 139:15: "My frame was not hidden from thee, when I was being made in secret, intricately wrought in the depths of the earth" bears traces of the mother-earth's womb in older creation myths.[5]

Many ancient pagan accounts of creation involve an act of dividing elements, e.g., light from darkness or land from water. For instance: Marduk, the god of light, divides the dragon Tiamat's corpse to produce the heavens and the earth in the Enuma Elish; the heavens are set at a distance from the earth in the Gilgamesh myth; and in the Egyptian narrative, Shu, the god of air, separates Geb, the god of earth, from Nut, the goddess of heaven.[6] This is not absent from the Bible, where we read that God separated "the light from the darkness," "the waters which were under the firmament from the waters which were above the firmament," and the water from the land.[7]

The presence of the idea of creating through the utterance of a word has been documented in ancient pagan myths. In Egyptian literature we even encounter the accompanying approval of the result: "every utterance of the god truly came into being through that which was conceived by the heart and commanded by the tongue.... Thus was Ptah satisfied, after he had made all things and every divine utterance." This theme is emphasized in Genesis 1, where we read eight times that God uttered a command like "let there be" as he created things during the six days; and six times, it is reported that God saw afterwards that "it was

---

5.　Westermann, *Genesis*, pp. 25–26.

6.　Ibid., pp. 33–34.

7.　Gen. 1:4, 7, 9.

good." However, dependence cannot be established since these accounts were composed at about the same time.[8]

The story of the formation of man out of clay seems to be the most common creation motif. As an illustration, in the Egyptian narrative, the potter god Chnum created humans out of clay using a potter's wheel.[9] In Gen. 2:7, we learn that "the LORD God formed man of dust from the ground, and breathed into his nostrils the breath of life."

The idea of man's being created in a god's image is not as old as the story of his being created out of clay, but the notion, though less explicit than in the Bible, does appear in the Epic of Gilgamesh, where Aruru creates Enkidu out of clay after conceiving a "double of Anu," the sky god. In the Bible the human person's dignity as son or daughter and covenantal partner of God is clearly affirmed over against the notion that man was created to do the work of junior gods that grew tired of it, such as one finds in Mesopotamian literature.[10] In Genesis, tiresome work is identified as a punishment, not man's essential purpose. Man tilled the Garden of Eden for himself, not for God's benefit.[11] And only after the Fall would he have to do so by toil and sweat, reaping thorns along with his food.[12]

The struggle motif that appears in ancient creation literature, such as Marduk's killing Tiamat, is reflected in the Bible in the prophets and wisdom literature. For example, we find the Lord doing battle with and conquering dragons, serpents, sea monsters, and watery chaos in: Job 26:12–13: "By his power he stilled

---

8. Ibid., pp. 38–39.
9. Ibid., p. 35.
10. Ibid., pp. 36–37.
11. Gen. 2:15.
12. Gen. 3:17–19.

the sea; by his understanding he smote Rahab. By his wind the heavens were made fair; his hand pierced the fleeing serpent"; Ps. 74:13–15: "Thou didst divide the sea by thy might; thou didst break the heads of the dragons on the waters. Thou didst crush the heads of Leviathan, thou didst give him as food for the creatures of the wilderness. Thou didst cleave open springs and brooks; thou didst dry up ever-flowing streams"; Ps. 89:10–11: "Thou dost rule the raging of the sea; when its waves rise, thou stillest them. Thou didst crush Rahab like a carcass, thou didst scatter thy enemies with thy mighty arm"; Is. 27:1: "In that day the LORD with his hard and great and strong sword will punish Leviathan the fleeing serpent, Leviathan the twisting serpent, and he will slay the dragon that is in the sea"; and Is. 51:9: "Was it not thou that didst cut Rahab in pieces, that didst pierce the dragon?"

Modern biblical scholars have determined that the names Rahab and Leviathan come from Ugaritic literature. Babylonian, Canaanite, and Egyptian influences are also detected. In the Bible, however, in most cases, the struggle is not linked with creation but with the people's lament.[13]

There is only the faintest hint of any struggle in Genesis, in the mention of chaos or formlessness: "The earth was without form and void, and darkness was upon the face of the deep."[14] In Babylonian myths, the struggle against chaos, already a theme in older Sumerian myths, became a central point in the creation story.[15] Whether Genesis manifests a correction of the Babylonian idea that the creator struggles or was simply not influenced by that narrative, the inspired author illustrates that the true God

---

13. Westermann, *Genesis*, pp. 32–33.
14. Gen. 1:2.
15. Westermann, *Genesis*, p. 31.

does whatever he wills effortlessly. Another theological purification of the Enuma Elish that Genesis offers is that man does not come from an evil seed, i.e., the dead dragon's blood, and he does not arise from strife.

In conclusion, while Israel shares similar themes in its religious poetry with its pagan neighbors—a fact we need not fear—there is a definitive new revelation that works as a corrective in all these narratives. Through some of these same ancient images, the sacred Scriptures teach anew that there is only one God, uncreated and omnipotent, the only source of all things, who has bestowed on man the great dignity of being made in the image of the Almighty.

John Baptist Ku, O.P.

❧ CHAPTER 19

Catholic Teaching on Creation and on Human Origins

Since divine revelation and scientific discovery are both gifts that come from God and that can guide us back to God, they cannot contradict each other. Hence, a genuinely Catholic appropriation of the discoveries and insights of evolutionary science would have to be consonant with God's revelation of Himself that was to culminate in the person and mission of the incarnate Word, Jesus Christ.

As we begin these next chapters on Catholic theology and evolution, it is important, therefore, that we contextualize our discussion within the framework of the Catholic Church's doctrine on creation and human origins. What has God told us about his creation not only of the world but also of the human creatures that he had made? What revealed truths would have to be brought into conversation with evolutionary theory?

To answer these questions, there is no better text to go to than the *Catechism of the Catholic Church,* which contains a catechesis that beautifully summarizes Catholic doctrine on creation. It speaks "first of the Creator, then of creation and finally of the fall into sin from which Jesus Christ, the Son of God, came to raise us up again."[1]

First, the *Catechism* teaches that the world is created. The ancient creeds of the Catholic Church confess that God the Father is "Creator of heaven and earth" (Apostles' Creed), "of all things visible and invisible" (Nicene Creed). The New Testament reveals that God created everything by the eternal Word, his beloved Son. It is through the Son that "all things were created, in heaven and on earth…all things were created through him and for him. He is before all things, and in him all things hold together."[2] The Church's faith likewise confesses the creative action of the Holy Spirit, the "giver of life", "the Creator Spirit," and the "source of every good."[3]

Metaphysically, this means not only that the eternal Triune God created the temporal world at the beginning of time, creating something where there was nothing, but also that he sustains the world in time, preventing what he has created from being annihilated, returning back to nothing. Nothing exists that does not owe its existence to God the Creator.[4]

Next, the *Catechism* teaches that the world was created for the glory of God. As St. Bonaventure explains, God created all things "not to increase his glory, but to show it forth and to communicate it."[5] God has no other reason for creating than his love and his goodness. Thus, it should not be surprising that he created an ordered and a good world. It is an ordered creation because it is the creation of a wise God, who wills the interdependence of all his creatures. Creatures exist only in dependence on each other, to complete each other, in the service of each other.[6]

---

1. *CCC*, §279.
2. Col. 1:16–17.
3. *CCC*, §291.
4. Ibid., §338.
5. See ibid., §293.
6. Ibid., §340.

Not surprisingly, it is also a beautiful creation where the order and harmony of the created world results from the diversity of beings and from the relationships that exist among them.[7] Lastly, it is a good creation because it shares in God's goodness, which is why the sacred Scriptures reveal that after he created the world, "God saw that it was good...very good."[8]

Third, the *Catechism* teaches that the creation of the human being was the summit of creation because he is made in the image of God. Of all the visible creatures, only the human being is able to know and to love his creator. He is "the only creature on earth that God has willed for its own sake."[9] God created everything for human beings, but they were in turn created to serve and to love God and to offer all creation back to Him.[10]

Significantly, sacred Scripture reveals that God created the first human beings in a state of harmony, not only with God, but also within themselves, with each other, and with all the other creatures around them. Our first parents were constituted in an original state of holiness so that they were able to share in divine life.[11] This original state, called the state of original justice, was a state of grace where the first human beings were free from the disordered tendencies that we experience today. As long as they remained in the divine intimacy, the first human beings did not have to suffer or to die.

Finally, however, the *Catechism* teaches that there was a historical fall when the original human beings fell into sin: "The account of the fall in Genesis 3 uses figurative language, but affirms a primeval event, a deed that took place at the beginning

---

7.   Ibid., §341.
8.   Gen. 1:4ff.
9.   *CCC*, §356.
10. Ibid., §358.
11. Ibid., §375.

of the history of man. Revelation gives us the certainty of faith that the whole of human history is marked by the original fault freely committed by our first parents."[12] The original sin involved disobeying God's command, and all subsequent sin would be disobedience toward God and lack of trust in his goodness.[13]

Sadly, because of this original sin, the harmony in which the first human beings had found themselves when they were created was destroyed. When they rejected God, they also rejected his gifts. This rejection led to disorder within the human being, between human beings, and between human beings and the other creatures that surrounded them. This privation of grace and of harmony was inherited by all subsequent human beings, who would therefore experience suffering and death. Christians, however, believe and are convinced that "the world has been established and kept in being by the Creator's love; has fallen into slavery to sin but has been set free by Christ, crucified and risen to break the power of the evil one."[14]

Nicanor Pier Giorgio Austriaco, O.P.

---

12. Ibid., §390.
13. Ibid., §397.
14. Ibid., §421.

❧ CHAPTER 20

The Web of Evidence for Evolution (I)

Like the theory of gravity, the theory of evolution is an explanation for a particular aspect of God's creation. Where the theory of gravity explains the relative motion of physical bodies with respect to each other—for instance the falling of an apple to the earth or the orbit of a planet around the Sun—the theory of evolution explains the history, diversity, and relationships of all living things with respect to each other, not only among those that are extant, i.e., still found in the world today, but also among those that have become extinct.

In brief, according to the theory of evolution, the history, diversity, and relationships among all living beings that have ever lived on our planet can be explained as follows: Life on our planet diversified gradually beginning with one primitive living thing that lived more than 3.5 billion years ago. Over time, this one living thing became many more living kinds of things via a mechanism that can be explained primarily (but not completely) by genetic change and natural selection. All living beings on our planet, human beings included, are descendants of a common ancestor.

What is the evidence for this explanation? Some believe that there is one definitive experiment that "proves" that evolution is true. However, like the theory of gravity, the theory of evolution is not based on any single observation. Rather it is supported by numerous observations from different areas of biological and paleontological research. Thus, the theory of evolution is justified by a web of evidence that together support the claim that all life on our planet has evolved from a common ancestor.

This web of evidence is comparable to the webs of evidence presented at a jury trial. It is not this piece of evidence alone—say the fingerprint of the suspect on the stolen item—but this piece of evidence along with all the other pieces of evidence—the video footage taken by the camera in the supermarket and the eyewitness account by another customer, for example—that convicts the suspect in an open-and-shut case. It is all the evidence taken together and the likelihood that they explain the crime better than any other alternative account that justifies the verdict.

The Fossil Record: First, the theory of evolution explains the fossil record well. It explains why the deepest and thus the oldest rocks in the fossil record contain the simplest fossils. It explains why later species in the fossil record—the fossils in younger rocks—have characteristics that make them look like the descendants of earlier ones. Lastly, it explains why transitional fossils—fossils of species that have characteristics that are typical of an older species and those that are typical of a later one—exist and existed at this particular place and time rather than at that particular place or time.

One of the most spectacular transitional fossils is the transitional species, *Tiktaalik roseae,* which was discovered in 2004 on Ellesmere Island in Nunavut, Canada. Until 390 million years ago (mya), the only vertebrates were fish. However, 30 million years later (360 mya), the earth had become populated by four-footed vertebrates that walked on land, called tetrapods. These early tetrapods resembled modern amphibians like frogs and salamanders in that they had flat heads and bodies, a distinct neck, and well-developed legs and limb girdles. However, they also resembled the earlier fishes in that they had scales, limb bones, and head bones.

The available evidence suggested that a transitional species—an extinct species that would have both fish-like and amphibian-like characteristics—would be found in rocks formed 375 mya between the 390 mya fish-only and the 360 mya amphibian-only rocks. This prediction was confirmed with the discovery of *Tiktaalik roseae* in rocks formed 375 mya, and not in rocks formed 400 mya or 350 mya.

Tiktaalik was an animal that had gills, scales, and fins that allowed it to live in water. In this way, it was fish-like. However, it also had amphibian-like traits including sturdy ribs that helped the animal to pump air into both its lungs and its gills. Interestingly, it also had limbs that are part fin, part leg, that allowed it to push itself up. It also had a neck, which fish that had skulls connected directly to their shoulders, did not have. Together, these traits suggested that *Tiktaalik* was adapted to live and to crawl in shallow waters, and to peer above the water surface and to breathe air.

The theory of evolution is the best explanation we have for the characteristics and the timing of the fossils found in the rocks of our planet.

Objection: Critics of evolutionary theory often raise several objections when confronted by the fossil record. Most prominently, they point to instances in the fossil record where new species appear "instantaneously" in the rocks.

The most striking example of this is the Cambrian explosion which began around 540 mya. Prior to the Cambrian explosion, most organisms were simple cell-like creatures that may have been organized into colonies. After the Cambrian explosion—a relatively short period of 70 or 80 million years—numerous complex multi-cellular organisms that are representative of all the major animal kinds today are found in the fossil record. Critics argue that evolutionary theory cannot explain the rapid and apparently unprecedented appearance of these animal kinds.

In response, it is clear that we still do not understand the Cambrian explosion completely. However, animal fossils from before the Cambrian, called the Ediacarans, have been found suggesting that the animals of the Cambrian did not appear without precedent. Moreover, a study of the Cambrian fossils of the arthropods, the animal kinds that include insects, spiders, and lobsters, shows that the Cambrian rates of evolution were within the boundaries of normal evolutionary processes.[1] In sum, this suggests that the Cambrian explosion is not beyond the explanatory power of the theory of evolution.

Nicanor Pier Giorgio Austriaco, O.P.

---

1.   Michael S.Y. Lee, Julien Soubrier, and Gregory D. Edgecombe, "Rates of Phenotypic and Genomic Evolution during the Cambrian Explosion," *Current Biology*, Vol. 23, No. 23 (2013): pp. 1889–1895.

❧ CHAPTER 21

The Web of Evidence for Evolution (II)

In the previous chapter, we considered the fossil record as one piece of evidence in support of the theory of evolution. Here, we discuss three additional threads of the web of evidence for evolution taken from organismal biology, from molecular biology, and from biogeography.

Organismal Biology: If evolutionary theory is true, then organisms should bear marks of their evolutionary history within their bodies. Biologists point to so-called vestigial organs as one of these marks. Vestigial organs are organs or structures in a living species that have apparently lost most or all of their ancestral function associated with the ancestral species. Their existence is best explained by the evolutionary history of the living species.

For example, whales and dolphins have a rudimentary pelvis or hipbone. Snakes and legless lizards have rudimentary pelvic bones as well. Why would these animals that lack limbs have a hipbone? Strikingly, whale, dolphin, snake, and legless lizard embryos initially develop embryonic hind legs or limb buds though these are reabsorbed by the developing organism before birth or hatching. Why do these animals have these transient rear limbs during their development?

The best explanation is that these creatures evolved from a common ancestor with four-limbed animals and that the vestigial hipbone and their vestigial embryonic limbs are marks of that evolutionary history.

In our species, it is striking that the human embryo initially develops a tail during his development in his mother's womb. At between four and five weeks of age, the human embryo has a dozen or so developing tail vertebrae which extend beyond his anus and his legs, accounting for more than 10% of the length of his body. By the ninth week of pregnancy, however, most of this extensive embryonic tail has undergone regression by a mechanism of programmed cell death. Why do we have these transient embryonic tails?

Once again, the best explanation is that we evolved from a primate ancestor that had a tail and that we retain a rudimentary genetic program for making tails. Incidentally, this would also explain those rare human beings who are born with an actual tail, which can be as long as 5 inches and which can move and contract. Here the normally silenced tail-making genetic program is reactivated and a rudimentary tail is made.

Molecular Biology: In recent years, biologists have been able to decode the information found in the DNA of numerous species of life—called their genomes—from bacteria to the pufferfish to the kangaroo to the human being. They have then been able to compare the genetic information of these diverse organisms to determine their similarities and differences.

Two discoveries stand out from this comparative analysis. First, all the genetic information in these diverse species are written in the same language, the same genetic code. Bacterial DNA, kangaroo DNA, and human DNA are all written with the same four chemical letters, abbreviated G, A, T, and C. Second, closely related species share more genetic information than distantly related ones. Thus, human beings share 96 percent of their genetic information with chimpanzees but only 75 percent of their genomes with pumpkins.

Why is this so? Evolutionary theory proposes that the common code and the similarity between DNA genomes of related species is a mark of their common ancestry.

To illustrate their reasoning, imagine that you are a high school teacher who is grading ten final exams. You discover that five of these exams are similar to each other, with similar word and phrase use, and that two of these five have identical answers. You therefore conclude that these exams have a common ancestry. It is likely that they are all descended from a single exam that was copied and shared among the cheaters, and that the two identical exams have a more recent ancestor in that they were probably copied, one from the other.

For the same reason, evolutionary biologists posit that organisms with similar genetic information have a common ancestor and that species with more similar genetic information have a more recent ancestor than those with dissimilar genomes. This is not a complex scientific argument. It is one based on ordinary everyday logic and reasoning.

Biogeography: Biologists have discovered that different locations on the planet with comparable geographical characteristics are actually populated with different kinds of organisms with comparable traits.

For example, desert flora have adapted to the extremes of heat and aridity found in those habitats. They are often succulents that have thickened and fleshy leaves and stems that are used to retain water. And yet different deserts have different types of succulents. In North and South America, the succulents are members of the cacti plant family, while in Asia, Africa, and Australia, the succulents are members of the euphorb plant family.

How do we explain this specific pattern in the global distribution of succulents?

Evolutionary theory proposes that this non-random distribution of succulents can be explained by the distinct evolutionary histories of the plants that evolved independently of each other in different but similar habitats on the planet.

In another example, mammals are found throughout the world in different geographical habitats. In most of the world, these habitats are populated by placental mammals that have placentas that allow their young to develop in their females. Flying squirrels, anteaters, and moles are examples of placental mammals. In contrast, in Australia, the parallel habitats are populated by marsupial mammals that have pouches and that give birth to very underdeveloped young. Supergliders, banded anteaters, and marsupial moles are the marsupial mammals that parallel their placental counterparts listed above. How do we explain this specific pattern in the global distribution of mammals?

Again, evolutionary theory proposes that this non-random distribution of animals can be explained by the distinct evolutionary histories of the organisms that evolved independently of each other in different but similar habitats on the planet.

In sum, the theory of evolution is supported by numerous observations from different areas of biological and paleontological research. It is justified by a web of evidence from the fossil record, from organismal biology, from molecular biology, and from biogeography, among others, that together support the claim that all life on our planet has evolved from a common ancestor.

I am often asked if evolution is a fact. For the scientist, the theory of evolution is as true as the theory of gravity. Is gravity a fact? If gravity is a fact, then evolution is a fact as well. The evidence is so overwhelming in favor of evolution that there really is no rival explanatory theory of repute for the origin and diversity of life on our planet, in the same way that there is no rival explanatory theory to the theory of gravity for the attraction of bodies.

Finally, I need to point out that evolutionary theory, as a science that seeks to explain the diversity and origin of life on this planet, has nothing to say about whether or not God exists. Evolutionary biologists who make this claim, and Richard Dawkins is one of them, do so not as scientists but as philosophers promoting an ideology that is sometimes called Neo-Darwinism. I accept the science of evolution. I reject the ideology of Neo-Darwinism.

Nicanor Pier Giorgio Austriaco, O.P.

❦ CHAPTER 22

The Fittingness of Evolutionary Creation

As a priest-scientist who supervises an NIH-funded research laboratory investigating the molecular regulation of cell death, I get a lot of science and religion questions from believers and non-believers alike. The second most common question I get—after the most common truth question, do you believe in evolution?—is the purpose question: Why did God choose to create via an evolutionary process rather than via special creation?

Many answers to this purpose question are possible, of course; but I have found that the most illuminating, and often the most surprising, response that I can give, is an argument based on the thought of St. Thomas Aquinas. It is a theological argument for the fittingness of evolutionary creation.

Aquinas was a Christian theologian whose most mature work, called the *Summa Theologiae*, remains a masterpiece of faith seeking understanding. In his writings, he frequently used theological arguments for fittingness to reveal the meaning, beauty, and wisdom of God's actions in the world. Arguing from fittingness involves understanding why an end is attained better and more conveniently with the choice of a particular means rather than another. In this sense, and as Aquinas himself explains, choosing to ride a horse is more fitting than walking if one seeks to quickly reach one's destination on a journey.[1] Theologically, arguments from fittingness try to explain how God's choice of a particular means allowed him to most appropriately attain the end of his actions.

It is important to acknowledge at the outset that theological arguments from fittingness are not demonstrative. In other words, they cannot prove that a certain conclusion necessarily has to be the way that it is. They cannot prove that the conclusion is true. It may be fitting for someone to ride a horse to reach his destination, but he may in fact have chosen to walk instead. Theological arguments from fittingness do not prove doctrine. They attempt to reveal the inner coherence and the wisdom of the divine design, the theo-drama that has been revealed by a God who is true, good, and beautiful.

Nonetheless, these arguments have been deployed by Christian theologians throughout the history of the Church to illustrate the coherence, the intelligibility, and the beauty of the Christian faith. For example, the author of the Letter of the Hebrews argues that it was "fitting that God, for whom and through whom all things exist, in bringing many children to glory, should make the pioneer of their salvation perfect through suffering."[2]

Returning to our purpose question in light of the Christian theological tradition, we can reword it as follows: Why was it fitting for God to have created via an evolutionary process rather than via special creation?

To answer this question, recall that for Aquinas, theological arguments from fittingness attempt to explain how God's choice of a particular means allowed him to most appropriately attain

---

1. See *ST* III, q. 1, a. 2.
2. Heb. 2:10.

the end of his actions. Therefore, to grasp my argument for the fittingness of God's creating via evolution, we need to begin by identifying the end of creation. Why did God create?

For the Catholic theological tradition, the answer to the purpose-of-creation question is clear: God chose to create because he wanted to manifest and to communicate his glory. *The Catechism of the Catholic Church*, the definitive summary of Catholic doctrine, proclaims that "Scripture and Tradition never cease to teach and celebrate this fundamental truth: 'The world was made for the glory of God.'"[3]

How does God communicate his glory to his creatures? According to Aquinas, God communicates his glory to his creatures by inviting them to participate in his existence. Creatures exist because God, whose essence is existence itself, gives them a share in his existence. This is the fundamental metaphysical distinction that distinguishes the Creator from his creatures: he has existence by nature, while they have existence by participation.

However, Aquinas also explains that God shares his perfections with his creatures by inviting them to participate in his causality, which in the world manifests itself in his governance of his creation:

> But since things which are governed should be brought to perfection by government, this government will be so much the better in the degree the things governed are brought to perfection. Now it is a greater perfection for a thing to be good in itself and also the cause of goodness in others, than only to be good in itself. Therefore God so governs things that he makes some of them to be causes of others in government; as a master, who not only

---

3. *CCC*, §293.

imparts knowledge to his pupils, but gives also the faculty of teaching others.[4]

To put it another way, according to Aquinas, it is a greater perfection, and therefore, more fitting, for God to share his causal power with his creatures, making them authentic causes that can cause by their own natures, than for God to remain the sole cause acting within creation.

As I have explained to my students at Providence College, it is easy for human beings to write a book, but it is impossible for them to make a book that writes itself. On the other hand, God not only causes, but also creates creatures who are in themselves, true causes. As such, when God does create creatures who themselves can cause, he manifests his power in a singular manner that signals his omnipotence.

Building upon this Thomistic theological account, I propose that it was fitting for God to have created via evolution rather than via special creation because in doing so, he was able to give his creation—the material universe and the individual creatures within it—a share in his causality to create. In this way, he more fully communicates his perfection to his creation, thus, more clearly manifesting his glory. As Aquinas points out: "If God governed alone, things would be deprived of the perfection of causality. Wherefore all that is effected by many would not be accomplished by one."[5]

Note that this is not the causality that allows one to create from nothing, because this causality is the sole prerogative of God who alone is creator. Rather, it is the causality that allows one to generate novelty and diversity from pre-existing matter.

---

4. *ST* I, q. 103, a. 6.
5. *ST* I, q. 103, a. 6.

This is also not the causality that philosophers call primary causality. Again this is the sole prerogative of God who is able to act solely on his own power. Instead, it is the causality called instrumental causality where God, the primary cause, activates the instrumental causality of his creatures so that he and they can act together wholly and fully, to create, in the same way that an author and his pen work together wholly and fully to write a letter. This is the kind of causality that underlies biological evolution.

In my view, at least three further points follow from this theological argument for the fittingness of evolutionary creation. First, I propose that once God had chosen to create through his creatures, it was fitting that he used evolution to create rather than another means, because evolution is the most efficient way for Divine Providence to use non-personal instrumental causes to generate novel and adaptive life forms on a dynamic and ever changing planet.

Take the Chicxulub asteroid strike that crashed into what is now the Yucatan Peninsula in Mexico approximately 66 million years ago. There is significant evidence that suggests that this asteroid strike, which left a 110-mile wide crater now buried nearly a mile underground, triggered the mass extinction at the Cretaceous-Paleogene boundary that killed off the dinosaurs. This mass extinction emptied ecological niches throughout the planet that could now be filled with novel plant and animal life.

In my view, evolution was the most efficient and fruitful way for God to use non-personal instrumental causes to create novel life forms after this planetary-wide extinction event, because a Darwinian evolutionary mechanism can shape and transform pre-existing life forms so that their surviving progeny can diversify and adapt to the increased number of available ecological niches. How else could God have used the non-personal instrumental causality of matter to create the novel kinds of mammals and birds that emerged to become the dominant land and marine vertebrates after the Chicxulub asteroid strike wiped out the dinosaurs?

Next, because of the fittingness of evolutionary creation, I also maintain that God did not "waste" life when he chose to create via an evolutionary process. This is a charge often levied against theistic evolution by creationists. For example, Henry Morris of the Institute for Creation Research argues that evolution cannot be reconciled with Christianity because "the standard concept of evolution involves the development of innumerable misfits and extinctions, useless and even harmful organisms. If this is God's 'method of creation,' it is strange that he would use such cruel, haphazard, inefficient, wasteful processes."[6]

In response, no one thinks that Michelangelo "wasted" marble because there were leftover marble pieces after he had completed sculpting his masterpiece, David. There is no waste when the agent fittingly attains his end. Likewise, I propose that extinct species are not pointless waste. Rather, they were the necessary "leftovers" from the creative evolutionary process that God used to generate the novel and diverse forms of life visible today in a manner most fitting to reveal his glory.

Finally, according to Aquinas, God created the diversity of creatures because no single creature can adequately reflect the perfection of God:

> We must say that the distinction and multitude of things come from the intention of the first agent, who is God. For he brought things into being in order that his

---

6. *Exploring the Evidence for Creation* (Eugene, OR: Harvest House Publishers, 2008), p. 172.

goodness might be communicated to creatures, and be represented by them; and because his goodness could not be adequately represented by one creature alone, he produced many and diverse creatures, that what was wanting to one in the representation of the divine goodness might be supplied by another. For goodness, which in God is simple and uniform, in creatures is manifold and divided and hence the whole universe together participates the divine goodness more perfectly, and represents it better than any single creature whatever.[7]

Therefore, in my view, it was also fitting that God created via evolution rather than via special creation because in doing so he was able to create more species to reflect his glory: Four billion species created over a three-billion-year period rather than just the eight million extant species today. In fact, it would have been ecologically impossible for all four billion species to co-exist on our planet, because there are only a limited number of ecological niches on the planet at a given moment in time.

To put it another way, there is a limit to the number of species and individual organisms that can be sustained by the planet at any one moment in time. Some of them are even mutually exclusive: If they had been created together, the large carnivorous dinosaur, *Tyrannosaurus rex*, would have wiped out the Asian elephant, *Elephas maximus*. However, with evolutionary creation—and not with special creation—these species were able to exist at separate moments in history to uniquely manifest the glory of their Creator. Again, they were not wasted.

To sum up, why did God choose to create via an evolutionary process rather than via special creation? Because it better

---

7.   *ST* I, q. 47, a. 1.

reveals his glory and his power. Because it reveals better that he is God.

Nicanor Pier Giorgio Austriaco, O.P.

❧ CHAPTER 23

How Does God Create Through Evolution?

So how does God create through evolution? How do we bring together an account of evolution with a Christian understanding of God's providential guidance of history? How can we think about God's working through a contingent series of events driven by mutation and natural selection? In a concrete way, how did God work to evolve the human FOXP2 which has been linked to our ability to speak and to understand a language?

Before discussing divine action, we should review some of the things we have already said about God, for as Aquinas and the other theologians of the Middle Ages understood, every agent only acts according to its nature. This is a commonsense principle: Cats do cat-actions like meow, kangaroos do kangaroo-actions like hop, and human beings do human being-specific actions like compose a symphony. To talk about God's activity, therefore, we first have to talk about God because only God can do God-actions. In doing this, we follow in the footsteps of Aquinas who, it is said, never ceased to pester his childhood teachers with the simple question: *What* is God?

As we have already discussed in earlier chapters on evolution and Christian faith, the passionate drive to answer this question—what is God?—propelled Aquinas to the heights of contemplative prayer and to a profound insight of theological brilliance: After carefully thinking about certain features of creation, he discovered that God must be unlike any other being in the world if he is to be its cause. He has to be radically different from everything that we see around us. Therefore, Aquinas concluded that God alone must be the act of existing itself. Or to put it another way, God's essence—what God is—is his existence. He is existing itself. This insight about the nature of God lies at the heart of Thomistic theology.

Understanding God to be the act of existing itself clarifies the distinction between the Creator and his creatures. As noted above, for Aquinas, God alone is existence by nature. In contrast, all other creatures are not God precisely because their act of existing has been received from God who alone is self-existing. In scholastic terminology, every creature has its existence not by nature but by participation.

To illustrate what Aquinas means when he says that creatures have their existence not by nature but by participation, he explains that a glowing rod of iron held in a fire is not fire by nature, but it is fire by participation. It is not fire, but it is like fire, because the true fire has made it like itself. In the same way, every creature exists because God who is existence itself holds every creature in existence at every time and place. If God ceased to hold us in existence, we would simply disappear. We would be annihilated.

Significantly, the distinction between the existence of the Creator by nature and the existence of his creatures by participation protects the integrity of both the Creator and his creatures. Within the Thomistic theological synthesis, the creature is able to have a distinct and, in a qualified sense, independent nature without severing its dependence upon its Creator.

To illustrate this claim, a kangaroo is a kangaroo because God made it a kangaroo. A kangaroo is not God because it has its own nature. In this sense, it is independent of God. However, a kangaroo's existence is existence by participation. It relies at every moment and in every place on God who makes it exist. In this second sense, it is radically dependent on God.

Consequently, because it possesses its own nature, a creature is able to act according to that nature without violating the Creator-creature relationship. In this way, the creature can claim some autonomy in its actions. Indeed, Aquinas argued that it is fitting and proper that God would create creatures that can act as true causes—in other words, that can act from their own natures like a cat meowing because it is a cat—since this better reflects his power. It manifests a greater benevolence to impart causality upon another rather than to withhold it.

To illustrate this claim using a crude analogy, it would be more fitting for God to make a book write itself rather than to write it Himself, since only he could accomplish the former while even human authors could accomplish the latter. Thus, for Aquinas, God is the primary cause giving existence to creatures who, acting according to the powers of their natures that God gives them, are true secondary causes. He can be compared to the manufacturer, the primary cause, of the fountain pens, which can be used to write as secondary causes.

With Aquinas' theology of creation in mind, we can now turn to a more in-depth discussion of how God creates through evolution. God acts in the world first as a first cause who gives things their existence. He also acts as a primary cause working through secondary created causes. In an analogous sense, God is like the carpenter who makes a puppet. It is only an analogy because God in creating the puppet gives the puppet its own nature and therefore its own independence in a way Geppetto could never have created Pinocchio. It is only God who is able to make a creature—say Geppetto the craftsman—and give the creature its own natural powers—Geppetto's ability to make a puppet.

Thus, divine causality because it is transcendent cannot be equated to the creaturely causality that we see around us. When God acts as an efficient cause in creating a creature, he does not "make it move" as creaturely agents like human engineers make the things that they "create" move. Rather God makes the creature movable by nature so that it can move itself. Further, as Creator, God is not simply the efficient cause for the existence and actions of his creatures. Since he gives each creature its nature, he is also responsible for its material, formal, and final causes.

First, by making a being a particular existing kind of thing here and now, say a rose bush, God specifies its formal and material causes. God makes it a rose bush and not a cherry blossom. Next, by giving a creature a nature which is ordered to a particular end, God specifies its final cause. He gives the rose bush a natural inclination to make roses and other rose bushes.

Thus, when a creature acts, God acts in and through it by sustaining it in existence, giving it the nature that is the source of its actions, and actualizing that nature in the here and the now. All of these facets of divine agency are only possible because God is the act of existing itself. God creates the rose bush by giving it its existence; he gives it its natural ability to make roses because he gives its nature its existence; and he actually moves it to make roses when its nature is inclined to do so in the early summer because he gives these biological activities their existence.

To illustrate this proposal in the context of biological history, consider the evolution of the human gene *FOXP2* which has been linked to our ability to speak and to understand a language. For the sake of discussion, let us say that the mutation which gave rise to the human gene that facilitated language use in our species occurred when a particular DNA repair molecule in a particular proto-human being who was anatomically human but who did not have the ability to speak, repaired a DNA strand damaged by high-energy radiation in a particular place and time in southern Africa.

According to the Thomistic account of divine agency described above, God acts in this event as first cause because he gives the DNA repair molecule and the DNA strand their existence as particular kinds of things with particular natures. The DNA strand can be repaired by the DNA polymerase because God made them what they are. Indeed, the DNA repair molecule was able to introduce a random mutation into the *FOXP2* gene precisely because God knew it and thus created it as error-prone and capable of randomly making mistakes. In introducing the genetic mutation into the DNA strand, the DNA repair molecule was functioning according to its God-given and God-guided nature. It was doing what it does as God had made it.

Moreover, the mutagenic event that made human *FOXP2* what it is today, can be said to have been ordained from all eternity, and in this sense be providential, because in knowing the DNA repair molecule as error-prone, God knows it as error-prone and existing at a particular time and in a particular place. Therefore, the contingent event which gave rise to human *FOXP2* occurred at the time and place that it did because God knew it and allowed it to exist precisely as happening at that time and place. It is a contingent event because God made it such.

But how can God know contingent events without undermining their contingency? For God, all created temporal events are present to him simultaneously in the eternal here and now. They remain contingent—they could have been otherwise—though they are known by God as necessary events happening now. Consider this analogy: When I see my student sitting in a chair in my classroom, his sitting in a chair remains contingent—he could have been standing up. However, I know this contingent event necessarily because when I see my student sitting in that chair at a particular time and place, he cannot be doing anything else right there and then.

In the same way, the eternal God knows all temporally contingent events necessarily because he sees them as happening right here and now. Indeed, they are contingent precisely because he knows them as such.

Evolution understood within the perspective of classical theism would consist of innumerable events where God working as First Cause determines, as only God can, in an analogous and uniquely divine sense, the course of every contingent event in evolutionary history. He would do this by working through the individual and contingent things, whether they are molecules, cells or organisms, that he has brought into existence at particular times and places, as particular kinds of things. God can design through chance.

Finally, I note that within this conceptual framework, the perceived problem of reconciling a changing world and a non-changing God who are in relationship with each other is a non-starter. The created order is an evolving one precisely because God who does not evolve, knows it as evolving and gives it existence precisely as such. Thus, there will always be a radical relationship of existential dependency of the creature on its Creator: The changing creature can only be changing because it was created as such by an unchanging Creator.

Nicanor Pier Giorgio Austriaco, O.P.

❧ CHAPTER 24

How Did God Create *Homo Sapiens* Through Evolution?

In previous chapters on evolution and Christian faith, I explained that an account of biological evolution can be reconciled with a robust understanding of God's providence because God acts as a transcendent cause who can work in and through the natures of his creatures. In this chapter, I would like to extend my remarks to the evolution of *Homo sapiens*. How did God create our species through evolution?

At this point, it is important to recall that human beings are spirit-matter composites. Each of us has a rational soul that informs a human body.

What exactly is a soul? Soul is how we explain life. When someone asks me why an apple falls to the ground, I tell him about gravity. Gravity explains why things fall. In the same way, when someone asks me why a cat is alive while a book is not, I tell him about soul.

Soul explains why things are alive. A cat has a soul while a book does not. Note that if soul is the explanation for life, then every living thing has soul: A living rose is informed by a rose soul; a living kangaroo is informed by a kangaroo soul; and a living human being is informed by a human soul.

Reflecting on the capacities of the human soul, Aristotle and Aquinas concluded that our soul, unlike the rose soul or the kangaroo soul, is immaterial. They thought that the human soul had to exceed the material because it was able to do things that exceed the capacities of matter. Specifically, they argued that the human soul is immaterial because it is able to grasp and process abstract ideas like "truth" or "justice" or "beauty" that do not have length, width, or height. To put it another way, the human soul is immaterial because it has the capacity to grasp the complex and abstract ideas that are presupposed by human language.

One of the things that Aristotle and Aquinas explained is that a soul is "fitted" to its body in the same way that a key is fitted to its lock. Thus, a human soul can only inform a body that is able to support those capacities. Biologically, this means that a human soul can only inform a body that has a brain that is complex enough to deal with and process language.

From a theological perspective, therefore, biological evolution was a 3.5 billion-year process, directed by God, to advance living matter until it was apt to be informed by a human soul. As we will see in a later chapter, this critical point in evolutionary history occurred 100,000 years in southern Africa among a group of anatomically modern human beings when an individual evolved the neurocognitive capacity for language.

As I explained in the previous chapters, evolution can be reconciled with a robust understanding of Divine Providence. The International Theological Commission of the Catholic Church, then chaired by Joseph Cardinal Ratzinger, put it this way:

> [A]ccording to the Catholic understanding of divine causality, true contingency in the created order is not incompatible with a purposeful divine providence. Divine causality and created causality radically differ in

kind and not only in degree. Thus, even the outcome
of a truly contingent natural process can nonetheless fall
within God's providential plan for creation. According
to St. Thomas Aquinas: "The effect of divine providence
is not only that things should happen somehow, but that
they should happen either by necessity or by contingency. Therefore, whatsoever divine providence ordains to
happen infallibly and of necessity happens infallibly and
of necessity; and that happens from contingency, which
the divine providence conceives to happen from contingency" (*ST* I, q. 22, a. 4 ad 1).[1]

This is why it is intelligible for the Catholic theologian to
claim that God guided the contingent process of evolution to
advance living matter until it could be informed by a human
soul. He guided biological history in the same way that he guides
human history. He does so without undermining the contingent
nature of history.

Importantly, only matter can evolve. Because it is immaterial, the human soul has to be created immediately by God. It
cannot evolve.

Finally, given my reliance on Aquinas, some may claim that
he would be opposed to this account of human evolution that
claims that matter evolved, because he thought that God created
all living things immediately at the beginning of time. This is not
accurate. It is clear that St. Thomas acknowledged that at least
one human being, in this case, Eve, existed materially before she
was created by God:

---

1. International Theological Commission, "Communion and Stewardship: Human Persons Created in the Image of God," §69. Available at
   http://www.vatican.va/roman_curia/congregations/cfaith/cti_documents/
   rc_con_cfaith_doc_20040723_communion-stewardship_en.html.

Nothing entirely new was afterwards made by God, but all things subsequently made had in a sense been made before in the work of the six days. Some things, indeed, had a previous experience materially, as the rib from the side of Adam out of which God formed Eve.[2]

Note that for St. Thomas, other species like the mule existed beforehand in their matter as well as in their causes:

Some things, indeed, had a previous experience materially, as the rib from the side of Adam out of which God formed Eve; whilst others existed not only in matter but also in their causes, as those individual creatures that are now generated existed in the first of their kind. Species, also, that are new, if any such appear, existed beforehand in various active powers; …. Again, animals of new kinds arise occasionally from the connection of individuals belonging to different species, as the mule is the offspring of an ass and a mare; but even these existed previously in their causes, in the works of the six days. Some also existed beforehand by way of similitude, as the souls now created.[3]

What is important is that the creation of each one of these new species has to refer back ultimately to God's creative activity. God creates through evolution.

Nicanor Pier Giorgio Austriaco, O.P.

---

2. *ST* I, q. 73, a. 1, ad 3.
3. Ibid.

CHAPTER 25

The Historicity of Adam and Eve
(I: Theological Data)

One of the most controversial disputed questions in the dialogue between evolution and Christian faith today involves the historicity of Adam and Eve. Did they *really* exist? Does it even matter to the Catholic faith whether they existed? Why or why not?

In the next several chapters, we will answer these questions as an exercise of faith and reason that seeks to be faithful to the Catholic dogmatic tradition. Here, we will begin by laying out the theological context that establishes the Church's teachings on the origins of our species. In the next chapter, we will summarize the doctrine of original sin because these dogmatic statements are at the heart of our discussion. In the third chapter, we will summarize the scientific data that supports the narrative for the evolution of *Homo sapiens*. Finally, in the fourth chapter, we will propose a conceptual synthesis that seeks to be faithful to both the theology and the science.

For Catholic theologians, the most recent statement of the magisterium of the Catholic Church on the historicity of Adam and Eve is the papal encyclical, *Humani Generis*, promulgated by Pope Pius XII in 1950. In that letter addressed to the bishops of the Catholic Church, the Holy Father taught the following:

When, however, there is question of another conjectural opinion, namely polygenism, the children of the Church by no means enjoy such liberty. For the faithful cannot embrace that opinion which maintains that either after Adam there existed on this earth true men who did not take their origin through natural generation from him as from the first parent of all, or that Adam represents a certain number of first parents.[1]

Some Catholic theologians and lay faithful take this papal statement as definitive magisterial teaching that affirms the historical existence of a single original couple from whom all human beings are descended. To put it another way, they think that this encyclical definitively rules out polygenism, which is the theological theory that human beings are descended from several original first couples. The theological theory that human beings are descended from a single original couple is called monogenism.

However, these same theologians and lay faithful often fail to consider the rest of the paragraph in the same encyclical where Pope Pius XII explains his reasoning for his conclusion that polygenism cannot be embraced by the Catholic Christian. The Holy Father taught:

Now it is in no way apparent how such an opinion can be reconciled with that which the sources of revealed truth and the documents of the Teaching Authority of the Church propose with regard to original sin, which proceeds from a sin actually committed by an individual

---

1.   *Humani Generis*, §37.

Adam and which, through generation, is passed on to all
and is in everyone as his own.[2]

In other words, at face value, Pope Pius XII ruled out polygenism because he could not imagine how an account of several original first couples could be reconciled with the Church's
teaching on original sin. As we will discuss in the chapters
that follow, this was not surprising because scientists in 1950
believed that the human race was descended from several original first non-human couples who were scattered throughout
the planet.

As we will also see, however, scientists today now think that
our species is descended from a population of hominins living
in the same geographical area. Therefore, in the chapter on the
historicity of Adam and Eve, I will propose that this contemporary scientific account on human origins can be reconciled
with the Church's teaching on original sin. In fact, I will go even
further by suggesting that we can still defend the historicity of
Adam as the single and first human being from whom all of us
are descended.

Significantly, notice that Pope Pius XII makes no mention
of the Genesis text in his encyclical, because for Catholics, the
disputed question over the historicity of Adam and Eve does not
involve a debate over whether the biblical text should be interpreted literally or not. As we have discussed in earlier chapters,
for the Catholic Christian, biblical interpretation is a work of
both faith and reason that seeks to read the sacred text in line
with all truth, theological and scientific, both of which have their
source in God. It is a task that is guided by the Holy Spirit who
continues to work within and through his Catholic Church.

2.   Ibid.

Finally, it is important to acknowledge that the International Theological Commission chaired by then Cardinal Joseph Ratzinger, now Pope Emeritus Benedict XVI, has published a theological statement on evolution that is open to polygenism. In its document, *Communion and Stewardship: Human Persons Created in the Image of God*, published in 2004, the Commission acknowledges that the scientific evidence points to a polygenic origin for our species:

> While the story of human origins is complex and subject to revision, physical anthropology and molecular biology combine to make a convincing case for the origin of the human species in Africa about 150,000 years ago in a humanoid population of common genetic lineage.[3]

We will discuss this scientific evidence in Chapter 27.

The Commission then makes the following theological claim:

> Catholic theology affirms that that the emergence of the first members of the human species (*whether as individuals or in populations*) represents an event that is not susceptible of a purely natural explanation and which can appropriately be attributed to divine intervention.[4]

This suggests that both monogenism and polygenism remain viable theological opinions for Catholic theologians seeking to be faithful to the doctrinal tradition.

Nicanor Pier Giorgio Austriaco, O.P.

---

3. *Communion and Stewardship*, §63.
4. Ibid., §73 (emphasis added).

❧ CHAPTER 26

The Historicity of Adam and Eve (II: The Doctrine of Original Sin)

In his book, *Christianity in Evolution: An Exploration,* Jesuit theologian and Catholic priest Jack Mahoney, S.J., has proposed that the truths of evolutionary biology have made the Catholic Church's traditional teachings on human origins obsolete: "I argue that with the acceptance of the evolutionary origin of humanity there is no longer a need or a place in Christian beliefs for the traditional doctrines of original sin, the Fall, and human concupiscence resulting from that sin."[1]

Mahoney is not alone in holding this view, and there are many other scholars, both Catholic and Protestant, who think that these traditional Christian doctrines, especially the doctrine of original sin, need to be jettisoned.

In this chapter, I respond to these theologians by arguing that the doctrine of original sin is an integral part of divine revelation that not only emerges from our understanding that God is good but also explains our lived experience of human brokenness. In this chapter and the chapters that follow on the historicity of Adam and Eve, I will also show that it is a doctrine that is not incompatible with an evolutionary account of creation.

What is the doctrine of original sin? It is an explanation from divine revelation for the lived experience of our brokenness. As my students at Providence College will quickly acknowledge, human beings are broken. We do things that we know will hurt us in the long run, and we do not do things that we know will make us happy, again in the long run. They echo the Apostle Paul, who declared: "I do not understand my own actions. For I do not do what I want, but I do the very thing I hate."[2] Why is this?

According to the *Catechism of the Catholic Church,* our God who is goodness Himself created everything good. He created them flawless so that they would reflect his infinite wisdom and goodness.[3]

For the same reason, according to the *Catechism,* our original parents were not only created good, they were also established in friendship with their Creator and in harmony with themselves and with each other and with creation around Him.[4] Theologians call this original state of harmony, the state of original justice. It is a state of original goodness that was given to our first parents by a God who is good.

The "mastery" over the world that God offered to the original human beings in the state of original justice was realized above all within themselves. Our first parents had mastery of self. Aquinas would explain this mastery of self as follows:

> We saw above that man was originally constituted by
> God in such a condition that his body was completely

---

1. Jack Mahoney, *Christianity in Evolution: An Exploration* (Washington, DC: Georgetown University Press, 2011), p. 71.
2. Rom. 7:15.
3. See *CCC*, §339.
4. See ibid., §374.

subject to his soul. Further, among the faculties of the soul, the lower powers were subject to reason without any rebelliousness, and man's reason itself was subject to God. In consequence of the perfect subjection of the body to the soul, no passion could arise in the body that would in any way conflict with the soul's dominion over the body. Therefore, neither death nor illness had any place in man. And from the subjection of the lower powers to reason there resulted in man complete peace of mind, for the human reason was troubled by no inordinate passions. Finally, owing to the submission of man's will to God, man referred all things to God as to his last end, and in this his justice and innocence consisted.[5]

In other words, God gave our first parents who were in the state of original justice the sanctifying grace that justified them and made them righteous. Our human parents were friends of God.

God's supernatural gift of sanctifying grace, according to Aquinas, was also accompanied by three other gifts, called the preternatural gifts, that were given to our first parents to perfect them by remedying their natural weaknesses.

First, human beings by their very nature as creatures made of spirit and matter are inherently corruptible. In other words, because we are made of two things, spirit and matter, we naturally are prone to their separation. There is nothing inherent to spirit and to matter that would keep them together forever. This separation of spirit and matter is called death. The gift of immortality was given to the human beings in the state of original justice to perfect this inherent corruptibility so that they would live forever.

---

5. *Compendium of Theology*, §186.

Second, human beings by their very nature are prone to interior disarray because what we are inclined to know, what we are inclined to choose, and what we are inclined to desire often do not coincide. We lust after pleasures that are often at odds with what we know are reasonable. The gift of integrity was given to our first parents to perfect this interior disorder so that they would be more apt to act well in grace to attain their holiness. This preternatural gift orders persons so that their reason is subject to God, their desires are subject to their reason, and their bodies are subject to their souls.

Third, human beings by our very nature are inherently limited in knowledge because we know things through contingent realities and we learn about them in a gradual fashion. In Aquinas' view, the gift of infused knowledge was given to our first parents to remedy this weakness. It would have included all truths, both natural and supernatural, necessary for human beings to direct their lives and the lives of others, ordering everything to God.

Finally, because they had the gifts of immortality, of integrity, and of infused knowledge, the original human beings, according to Aquinas, were also impassable. In other words, they were not able to experience bodily or spiritual suffering.

Building upon this Thomistic theological account, I have also proposed that it would have also been fitting for God to have given the first human beings, several gifts that I have called the preteradaptive gifts as soon as they had evolved to perfect them not only as persons made of body and soul but also as persons who have evolved from non-personal primate ancestors.[6]

---

6. Nicanor Austriaco, O.P., "A Theological Fittingness Argument for the Historicity of the Fall of *Homo Sapiens*," *Nova et Vetera*, Vol. 13, No. 3 (2015): pp. 651–668.

These preteradaptive gifts would include gifts, among others, to counter the evolved adaptations we inherited from our primate ancestors, to infidelity, to violence, and to biased and false knowledge. These gifts would have given the first human beings the capacity to love faithfully, to peace, and to know truth.

In sum, in the state of original justice, our original parents would have been given everything that they would have needed to live in a state of harmony with themselves, with each other, with all of creation, and most significantly, with God their Father who created them.

And yet, we are broken! We are inclined to actions and to inactions that hinder us from the happiness that God had intended to give us from the very beginning. Why is this?

According to the *Catechism*, our existential brokenness can be explained by positing a historical event in the distant past when our first parents rejected God and all of his gifts. The Bible reveals that the original human beings were put to a test, a test that they failed through an act of disobedience. This was the original sin.

Some may wonder, why did God even test our first parents? Aquinas explains that God gave our original parents an explicit command that they had to obey so that they would learn from the very beginning to follow God's will.

When our first parents disobeyed God with the original sin, they rejected him and with Him, all of his gifts. Without the preternatural and the preteradaptive gifts, these fallen human beings thereafter became subject to death, to suffering, to error-filled knowledge, to weakness of will, and to disordered desire. In other words, they became subject to the interior struggle that is the source of our brokenness.

Notice that this account of original sin sees the effect of original sin, not as an addition to or a corruption of human nature—certainly not as a tendency to evil or a perversion that makes the human being evil as such—but as a privation of that nature, an absence, a lack, a wound, that leaves human beings struggling with the consequences of their nature as it had been created and evolved. It is this struggle that makes the human person prone to evil acts though he himself is not inherently evil.

Significantly, the Catholic Church teaches, as Aquinas had explained, that God had intended our first parents to give their descendants the blessings of original justice. When they lost the gifts, however, they could not give it to their posterity. Thus, the Council of Trent (1546) would teach that original sin is transmitted by propagation and not by imitation.

Note that some may read this statement by the Council of Trent—that original sin is transmitted by propagation and not by imitation—as a claim that original sin is transmitted biologically. However, the consequences of original sin are privations in the soul of the human being. Because of the original sin, his soul lacks grace and the preternatural gifts. Thus, when the Council of Trent teaches that the consequences of the original sin—our fallen human nature—are transmitted by propagation and not by imitation, we should understand this as a metaphysical and not a biological claim. As immaterial spirit, our soul and its properties are not determined by our genes. Instead, when God creates our souls when we are conceived, he creates them without the graces and gifts that we should have inherited from our original parents.

In conclusion, to summarize the Catholic Church's teaching on original sin, the *Catechism* puts it as follows:

> The transmission of original sin is a mystery that we cannot fully understand. But we do know by Revelation that Adam had received original holiness and justice not for himself alone, but for all human nature. By yielding to

the tempter, Adam and Eve committed a personal sin, but this sin affected the human nature that they would then transmit in a fallen state. It is a sin which will be transmitted by propagation to all mankind, that is, by the transmission of a human nature deprived of original holiness and justice. And that is why original sin is called "sin" only in an analogical sense: it is a sin "contracted" and not "committed"—a state and not an act.[7]

Thus, the doctrine of original sin explains why we are broken. However, it is also the grounds for our salvation because "after his fall, man was not abandoned by God."[8] Thus, it should not be surprising that the *Catechism* concludes: "The Church, which has the mind of Christ, knows very well that we cannot tamper with the revelation of original sin without undermining the mystery of Christ."[9]

Nicanor Pier Giorgio Austriaco, O.P.

---

7. *CCC*, §404.
8. See ibid., §410.
9. Ibid., §389.

❧ CHAPTER 27

The Historicity of Adam and Eve (III: Scientific Data)

In Chapter 25, I summarized the theological data that establishes the Catholic Church's teachings on the origins of the human race. A believer seeking to be faithful to the Catholic tradition and to divine revelation would need to take this theological data into account as he or she strives to respond to concerns raised by theologians who question the historicity of Adam and Eve.

In this chapter, I will summarize the scientific data that grounds the scientific explanation for the origins of our species, *Homo sapiens*. Basically, I will outline the evidence for what biologists call the Out-of-Africa model for human origins. In Chapter 28, I will then propose a theological narrative, an exercise of both faith and reason, that synthesizes the theological and scientific data into a coherent whole.

In 1950, when Pope Pius XII published his encyclical, *Humani Generis*, scientists thought that the human race had evolved independently from different non-human populations that had existed in different regions in the world before the appearance of *Homo sapiens*. These archaic non-human populations were members of the species, *Homo erectus*.

According to this Multi-Regional Model for human origins, native Africans had evolved from archaic non-humans in Africa, native Europeans had evolved from archaic non-humans in Europe, native Asians had evolved from archaic non-humans in Asia, and native Australians had evolved from archaic non-humans in Australasia. Though there was some interbreeding between these evolving human populations over the course of history, according to this explanation for human origins, the peoples of Africa, of Europe, of Asia, and of Australasia evolved into modern human beings relatively independently.

From a theological perspective, according to this multi-regional model for human origins, it is hard to see how every human being who ever lived could have shared a common fallen human nature that would eventually be assumed and redeemed by Christ. For example, if the historic event of the fall had taken place in Africa, how could this event have affected the human natures of the individuals who had evolved independently in Australia, by propagation and not by imitation? It is not clear. Therefore, as we saw in Chapter 25, it is not surprising that Pope Pius XII had taught that,

> it is in no way apparent how [polygenism] can be reconciled with that which the sources of revealed truth and the documents of the Teaching Authority of the Church propose with regard to original sin, which proceeds from a sin actually committed by an individual Adam and which, through generation, is passed on to all and is in everyone as his own.[1]

With the Multi-Regional Model, there could never have been an original Adam or an original Eve.

---

1.   *Humani Generis*, §37.

By 2010s, the science had changed dramatically. Today there is robust evidence from both the fossil record and genetic data that anatomically modern humans—creatures that looked liked us—evolved in Africa between 200,000 and 150,000 years ago, and that they migrated out of Africa about 60,000 years ago.[2]

According to this Out-of-Africa Model, anatomically modern humans evolved about 150,000 to 200,000 years ago in southern Africa, and then about 100,000 years ago spread to other regions of Africa. Several tens of thousands of years later, a small group of humans exited northeastern Africa and continued this expansion throughout Europe, Asia, Oceania, and eventually the Americas.

Notably, anatomically modern humans lived alongside human-like, yet still non-human bipedal species, creatures called archaic hominins by scientists, including the Neanderthals and the Denisovans, who have since gone extinct. There is strong evidence that these species interbred with our own such that 1 percent to 4 percent of the DNA of human beings living today who are not of African ethnic descent is of Neanderthal origin, and between 3 percent to 5 percent of the DNA of Melanesians and Aboriginal Australians is from the Denisovans.

How many original human beings were there? Studies suggest that the ancestral effective population size for anatomically modern humans in Africa is about 10,000 breeding individuals. In other words, one would need to posit the existence of 10,000 original humans to properly account for the genetic diversity that we see among the seven billion human beings living today.

Note that it is unlikely that these original humans lived in the same community, because without agriculture—which only

---

2. Brenna M. Henn, L.L. Cavalli-Sforza, and Marcus W. Feldman, "The great human expansion," *Proceedings of the National Academy of Sciences USA*, Vol. 109, No. 44 (2012): pp. 17758–17764.

appeared about 14,000 years ago—they would not have been able to find enough food to support a single tribe larger than several hundred members.

At this point, however, I need to make a critically important distinction between anatomically modern humans—bipeds who looked liked us—and behaviorally modern humans—bipeds that not only looked like us, but also behaved like us as well. Though anatomically modern humans evolved around 200,000 to 150,000 years ago, behaviorally modern humans did not appear until much later.

This critical transformation from anatomically modern to behaviorally modern human beings—called the Great Leap Forward by evolutionary biologist Jared Diamond[3]—is revealed by a rich archeological record of painting, engraving, carving, bodily decoration, notation and music. The earliest archaeological evidence for such modern behavior has been linked to artifacts found in Blombos Cave in modern-day South Africa dating to about 75,000 years ago.

Biologically, this transformation from anatomically modern to behaviorally modern human beings is often attributed to the evolution of brain structures that would have facilitated the use of language. Ian Tattersall, Curator Emeritus of the American Museum of Natural History has noted, "If we are seeking a single cultural releasing factor that opened the way to symbolic cognition, the invention of language is the most obvious candidate. Indeed, it is perhaps the only plausible one that it has so far proved possible to identify."[4]

3.　Jared Diamond, *Guns, Germs, and Steel* (New York: W.W. Norton & Company, 2005), pp. 39ff.

4.　Ian Tattersall, "How We Came to Be Human," *Scientific American—Special Edition: Becoming Human*, Vol. 16, No. 2 (2006): pp. 66-73.

Strikingly, Noam Chomsky, whom many consider the father of modern linguistics, has proposed that this cognitive transformation that changed a non-linguistic to a linguistic primate occurred in a single individual:

> It looks as if—given the time involved—there was a sudden "great leap forward." Some small genetic modification somehow that rewired the brain slightly [and] made this human capacity [for language] available.... Mutations take place in a person, not in a group. We know, incidentally, that this was a very small breeding group—some little group of hominids in some corner of Africa, apparently. Somewhere in that group, some small mutation took place, leading to the great leap forward. It had to have happened in a single person.[5]

For Chomsky and his colleagues, the human capacity for language is best explained if it first appeared in one individual – the first behaviorally modern human – who himself was one member of the larger population of anatomically modern humans. Interestingly, there is also data that suggests that all human languages are derived from a single proto-language that dates to about 100,000 years ago in central and southern Africa, though this claim is controversial.[6]

Finally, why is this transformation from anatomically modern to behaviorally modern humans so important for our discussion of the historicity of Adam and Eve? It is critically important

---

5. Noam Chomsky, *The Science of Language: Interviews with James McGilvray* (Cambridge: Cambridge University Press, 2012), p. 14.

6. Quentin D. Atkinson, "Phonemic Diversity Supports a Serial Founder Effect Model of Language Expansion from Africa," *Science*, Vol. 332, No. 6027 (2011): 346–349.

because, philosophically, this transformation can be understood to be archeological evidence for the appearance of the rational soul in human evolution. Theologically, this transformation would be a sign of the arrival on the stage of world history of the *imago Dei*, the creature made in the image and likeness of God with intellect and with will.

Nicanor Pier Giorgio Austriaco, O.P.

❧ CHAPTER 28

The Historicity of Adam and Eve
(IV: A Theological Synthesis)

In the previous three chapters, we examined the theological and the scientific data that need to be reconciled by faith and reason if we are to remain faithful both to divine revelation and to authentic scientific inquiry. In this chapter, I propose a theological narrative that seeks to do precisely this by bringing together themes that we have explored throughout this book. Note that this narrative remains only a hypothetical one that attempts to reconcile the data of faith and reason into a coherent whole.

From all eternity, the Triune God, Father, Son, and Holy Spirit, desired to share their life with persons who were not God. As such, they chose to create angelic and human persons who were destined to become like God by participation in the divine nature.

As pure spirits, the angelic creatures were created immediately. At their creation, some of them chose for God while others chose against Him. The former we call the angels; the latter we call the demons. As spirit-matter composites, the human creatures were created and are still being created over time. It was fitting for God, as we discussed in an earlier chapter, to create not only human beings but also all living material beings through an evolutionary process that better revealed his glory. It was then and remains now a process that is moved and directed by Divine Providence.

From a theologian's perspective, biological evolution was a 3.5 billion-year process, directed by God, to advance living matter until it was apt to receive a rational soul. This critical point in evolutionary history occurred about 100,000 years ago in Africa among a group of anatomically modern human beings when an individual hominin was conceived with the inherent neurocognitive capacity for language.

How exactly this happened will always be a matter of speculation. As Noam Chomsky has proposed, a single mutation acquired at conception could have altered the structure of an individual anatomical modern human's brain in a way that gave him the capacity for language. In my view, for reasons I cannot develop at this time, this capacity for language would have gone hand in hand with the capacity for abstract thought. As the first speaking primate, this individual would have also been the first rational animal. He was the first anatomically modern human to have the capacity to form abstract concepts, to reason, and therefore, to construct an internal map of his world. As he matured, he would have used this linguistic capacity to speak to himself and to God. This first speaking human is the original human we call Adam.

Moreover, since every human being today possesses the same linguistic capacity, each one of us must have inherited that capacity from him.[1] Each one of us must therefore be descended from this first speaking human. Adam would not only have been the

---

1. It is known that it is statistically very unlikely, if not impossible, for novel mutations of this kind to appear more than once in a hundred thousand-year period.

first speaking human, he is also the father of all speaking humans, which is all of us.

As we discussed in Chapter 26, it would have also been fitting for God to have given this original speaking biped—this behaviorally modern human named Adam—the supernatural graces and the preternatural gifts that he would have needed to attain his destiny to share the life of the Triune God. Adam would have been conceived in a state of original justice with the preternatural gifts that allowed him to order his instincts to his eternal destiny. Tragically, we know that this original human being ultimately disobeyed God in some way, forfeiting these supernatural and preternatural gifts he had been given, not only for himself, but also for his progeny.

Nonetheless, in the fullness of time, this original speaking biped and all of his descendants would be redeemed by Jesus Christ, the Savior of the World, so that they would be able, once again, to share in the inner life of the Trinity, and so, to live forever.

I am often asked three questions in response to this theological narrative. First, what about Eve? What can we say about her? In response, science can only speak about an original human individual who acquired a mutation that gave this person the linguistic capacity. Its interrogatory power is limited. Science cannot even tell us if that original human was a man or a woman. In this narrative, I have relied on the revelation of sacred Scripture to propose that the first speaking human was male. He was Adam. In the end, that this first speaking human was male and not female is a theological and not a scientific claim. It is a proposal of both faith and reason.

Furthermore, though science can affirm that there was an original human and that we are all descended from this person, it cannot tell us more about *how* that happened. There are many possible scenarios. All of them cannot be confirmed one way or the other by the scientific enterprise.

It could have been that God providentially ordered evolution so that two hominins, a male and a female, were conceived in the same community and at the same historical time with the mutations to be linguistic creatures. This is the option that I prefer because it is the most theologically fitting of all the scenarios I describe here. Significantly, it preserves the equal dignity of the man and the woman revealed in sacred Scripture since both of them were created at the same time and at the same place. It could also have been that God created Eve *from* Adam as the scriptural text describes. Both of these scenarios would preserve the historical narrative of an original human couple. They would have had children all of whom were speaking primates. They would have had children who were all human just like us.

It could also have been that God allowed Adam to mate with a non-speaking human female, with whom he had children. Some of these children would have been able to speak while others would not have been. Some of their children would have been human like us, while the non-speaking offspring would not have been. The children who spoke to each other would have mated among themselves giving rise to more speaking children.

Nonetheless, with all the possible hypothetical scenarios described above, given the advantage of speech that allow us to cooperate and to organize ourselves, the speaking individuals who would have been born would have come to dominate the entire population of *Homo sapiens*. In technical biological terms, the mutation that gave Adam the linguistic capacity would have become fixed in the human population.

Second, does this account not endorse sibling incest? This problem is not a new one. As Aquinas himself recognized, any theological account of a single original couple would have entailed sibling marriage to ensure the survival of the human race. Thus, he acknowledges that only parent-child relationships are excluded by the natural law. Brother-sibling relationships, though excluded today by law, would have been necessary early in the history of our species.[2]

Third, how should we understand the interbreeding that took place between behaviorally modern humans and their archaic hominin contemporaries, the Neanderthals and the Denisovans? This would have been matings between individuals who were both biological human but not both human persons like us. Theologically understood, these would be instances of bestiality of a sort. However, because of the similarities in appearance and in behavior among these closely-related hominin species, again both of whom are biologically but not personally human, the genetic similarity would have also made these matings fruitful in a way not possible today. In my view, the taboo against bestiality may not apply here precisely because of their biological compatibility.

Nicanor Pier Giorgio Austriaco, O.P.

---

2. See *ST* III, q. 54, a. 4.

CHAPTER 29

A Thomistic Response to the Intelligent Design Proposal

Intelligent Design (ID) is the proposal that "certain features of the universe and of living things are best explained by an intelligent cause, and not an undirected process such as natural selection."[1] It is a proposal that is linked to an intellectual, social, and political movement centered in and driven by a non-profit think tank called the Discovery Institute, which is based in Seattle, Washington, in the United States.

Recall that evolutionary theory proposes that organisms and living systems evolved in a step-wise and gradual manner. Pointing to bacterial flagella—the whip-like propeller structures used by bacterial cells to impel themselves through their microscopic environment—ID theorists propose that evolution cannot explain the gradual appearance of a bacterial flagellum and other molecular machines like it, because of, what they call, their "irreducible complexity."

ID proponent and biochemist Michael Behe explains, "By irreducibly complex, I mean a single system composed of several well-matched, interacting parts that contribute to the basic function, wherein the removal of any one of the parts causes the system to effectively cease functioning."[2]

An example of an irreducibly complex (IC) system according to Behe is a mousetrap. A simple mousetrap has no ability to trap a mouse until several separate parts—the base, the hammer, the spring, and the catch—are all assembled. This is what makes it irreducibly complex. According to ID proponents, however, a mousetrap because it is IC could not have evolved gradually: Intermediate structures in the evolution of a mousetrap, say a base with a hammer without a spring or a catch, would be non-functional as a mousetrap and therefore could not be subject to natural selection. Thus, a mousetrap must come to be at once in order for it to be functional.

This ID proposal for the non-evolvability of IC systems has been countered by evolutionary biologists in two ways. First, they have challenged the soundness of the argument for irreducible complexity. They have argued that sub-parts of an irreducibly complex system do not need to have the same function as the final system of which they are a part in order for them to have evolved. They need only serve some function—any function—in the cell.

Take the bacterial flagellum. The molecular structure of the bacterial flagellum suggests that it evolved from components of a bacterial pump. Yes, a sub-part of a flagellum could not function as a flagellum as ID proponents correctly point out, but evolutionary biologists counter that these sub-parts could have had another function. They could have been and were probably pumps. As such, these sub-parts could still be subject to natural selection. They could have evolved.

---

1. "Questions About Intelligent Design," Discovery Institute, Seattle, WA. Available at http://www.discovery.org/id/faqs/#questionsAboutIntelligent-Design. Last accessed on August 5, 2019.

2. Michael Behe, *Darwin's Black Box: The Biochemical Challenge to Evolution* (New York: Free Press, 1996), p. 39.

Significantly, this evolutionary mechanism would explain why the bacterial flagellum is a hollow tube rather than a solid whip-like structure, even though a whip would have made a better propeller than a straw. The bacterial flagellum is hollow because it evolved from a pump that is hollow.

Second, evolutionary biologists point out that the absence of known evolutionary pathways describing the appearance for the molecular machines described by ID proponents does not mean that they did not evolve. It is simply a sign of the incompleteness of science. In time, scientific research should uncover these pathways because, in principle, apparently IC systems like the flagellum are still subject to natural selection and evolution because their sub-parts could have original functions unlinked to their current use.

To add to this critique, I have also argued that ID proponents have neglected those apparent IC systems with components that come from two independent organisms.[3] Take the molecular machine used by the HIV virus to infect a human cell. This molecular machine, this infectivity structure, is made up of several components called gp120, gp41, CD4, and CCR5, that come together to form the lock-and-key structure needed for infection.

Note that this infectivity structure fits the definition of irreducibly complexity because all the molecular parts are needed together for viral infection, and loss of one part would destroy this function. And yet, gp120 and gp41 are viral proteins, and CD4 and CCR5 are human proteins! This is proof that living systems that fit the definition of IC could have evolved separately and independently.

---

3. Nicanor Austriaco, O.P., "The Intelligibility of Intelligent Design," *Angelicum*, Vol. 86, No. 1 (2009): pp. 103–111.

Finally, the ID proposal also raises other difficult questions, especially for theologians. Take the HIV infectivity structure that fits the definition of IC. There is convincing evidence that HIV first appeared in the 1930s in East Africa. If the ID proposal were true, this would suggest that the molecular machine for infectivity was "designed" less than one hundred years ago.

Did the designer creatively introduce the genes for CD4 and CCR5 in the human species in the distant past in anticipation of his introducing the genes for gp120 and gp41 in the HIV viral species in 1930?

Theologically, and more significantly, could this have been done by any other intelligent designer other than the intelligent designer commonly known as God? And if so, what does this say about God: Did he intentionally create us thousands of years ago so that we could be infected by a killer virus in the 20th century?

Why would a good and gracious Father who would not give his sons and daughters a snake if they asked for fish handicap his children from the start so that they would struggle and suffer later? These are just a few of the many profound and difficult—and I would add, unnecessary—theological questions raised by the ID proposal.

In the end, from the Thomistic perspective, the ID proposal is a misguided distraction. ID proponents claim that irreducible complexity is a sign for intelligent design. It is a sign of God's creative hand. However, like their counterparts pushing an atheist and a naturalist account of the world, they mistakenly assume that God works in his creation primarily by pushing and pulling atoms and molecules like a force, generally, and by assembling and disassembling living systems, more specifically. This god of ID is a small and puny god!

As we discussed in Chapter 27, the Creator God revealed in the sacred Scriptures creates through evolution primarily by giving existence to his creatures as individual members of a natural kind with specific capacities directed to a final end. Existence and not irreducible complexity is the sign *par excellence* for God's work! That these molecular machines even exist with their capacities directed towards a final end: This is the sure sign that they were created by an intelligent *and divine* Designer.

Nicanor Pier Giorgio Austriaco, O.P.

❧ CHAPTER 30

Evolution's End: The Beatific Vision

One of the reasons why many Christians are uncomfortable with the idea of human evolution is that it seems to suggest that the fact that we are here is a mere accident, and that there is nothing stopping us from evolving into something else down the road. There are many ways to address the question of why the human species evolved, but I want to focus on the end of evolution. Specifically, I want to ask the questions: Is there some stopping point for evolution? Is there a goal for the evolutionary process?

In looking for an end of evolution, we must first ask whether evolution has actually come to rest at any point in the past. For any plant or animal species we find in nature the answer is, in part, yes. Evolution has contributed to the production of a population of organisms that are, for the most part, adapted to their environment. Of course, this assumes that the environment is stable, which brings us to the "no" part of the answer. Any stable population of organisms that finds itself in a new environment, assuming it can survive at all, will begin to adapt to the new surroundings.

Like so many natural processes, therefore, evolution does not come with a determined endpoint. Its operation depends on many factors. Gravity may bring a boulder to rest on the edge of a cliff, but an additional push, whether by wind, earthquake, or human hand, will start the process of falling again until gravity finds the boulder a new place of rest. Arguing for an absolute endpoint intrinsic to evolution would seem to require that all evolution be pushing life to the same state of rest which, if the diversity of life is any indication, seems unlikely.

If there is no absolute endpoint for evolution, we can still ask about the specific states of rest we find. In particular, we might want to ask, is the human species still evolving? This is a complicated and much debated question. For instance, many cultural variations, ranging from skin color, to lactose intolerance, to the ability to breathe at higher altitudes, have been traced to particular populations of humans settling in particular geographic locations with their unique environmental pressures. Further, there is robust evidence for relatively recent adaptations among humans in our genetic resistance to various diseases.

It is clear that evolution is always at work on small scales, adapting the human species to better survive in particular environments. Nevertheless, there is no evidence to suggest that any one sub-group of the human population is diverging from the other individuals on the planet such that they would become a new post-human species.

Some have argued that our ability to change our environment—as we did with the invention of agriculture—and the fact that the human population is no longer geographically isolated have largely slowed the evolutionary forces that theoretically could have led to the genesis of several post-human species. Could we conceive of a scenario where we would not be able to control our environment enough to prevent the diversification of the human population into novel species? Perhaps, but at this time, that possibility seems to be thoroughly in the realm of science fiction.

What then of the second sense of end, the sense of a goal towards which evolution tends? Naturally speaking, because there is no absolute endpoint for evolution, as discussed above, there cannot be an absolute goal for this process of change and adaptation. There are only relative endpoints and relative goals involving the adaptation of this particular population in this particular environment at this particular time. And one of the relative endpoints for evolution, an endpoint that appears to be particularly stable, is the evolution of our species, a species that is able to radically control and alter the environment for its own sake and for its own survival.

But the scientific answer to the question of evolution's end is not the only answer available to faith and reason. Evolution, like all natural processes, is an instrument of God, caused and maintained in all of its working by his divine providence. Like all other instruments, the mechanism of evolution can be ordered to some end other than the one that it could attain by its nature alone.

Consider the craftsman who is able to transform a piece of wood into a violin. Wood on its own is not ordered towards the production of music, but in the hands of a master artisan and of an expert musician, it can do so much more than it could have done by nature. Indeed, from the perspective of a violin maker, the true "end" of a spruce tree is the masterful performance of a Beethoven symphony, though this end for the tree is not something that could have been discovered by the scientific method.

In the same way, as we explained in an earlier essay, from the theological perspective, we can affirm that biological evolution was a 3.5-billion-year process, directed by God, to advance living matter until it was apt to receive a human soul. By its nature, evolution is not intrinsically ordered to the appearance of an animal materially capable of being informed by an immaterial soul, but this is, in fact, what it has achieved through the ordering and providential hand of God, the master Creator.

Given that God has used evolution for the production of the human body, we can ask whether this particular goal of the divine plan is absolute, or whether it is a mere a stepping-stone to something else. Once again, our answer must be both yes and no.

In the Incarnation, Jesus Christ took to Himself our human nature in order to save us. Indeed, our nature has been redeemed, because it was assumed. In a letter, St. Gregory of Nazianzus explained this Catholic intuition this way: "For that which he has not assumed he has not healed; but that which is united to his Godhead is also saved."[1] If some sub-group of the human population evolved into a post-human species with a post-human nature, it is unclear how these individuals could and would share in the salvation merited for us by the Savior.

Thus, it seems unlikely that God would allow evolution to produce a post-human creature incapable of attaining beatitude because of his separation, in his very nature, from the humanity of Christ. It suggests that in the divine plan for salvation history, the appearance of the human creature is arguably the goal of evolution. While Aquinas knew nothing of biological evolution and could not speculate on its possible goal, he does speculate on the end on other natural processes, including the seemingly everlasting motion of the heavenly bodies. There he is perfectly willing to state, "We hold then that the movement of the heavens is for the completion of the number of the elect. ... It is a definite number of souls that is the end of the heavenly movement: and when this is reached the movement will cease."[2]

---

1. Epistle 101, 4th century A.D.
2. *De potentia*, 5.5.

With Aquinas, we can view biological evolution as ordered not only to the population of human beings in this world, but also to the population of human beings in the "the resurrection of the dead and the life of the world to come." This new life, begun in the sharing of the inner life of the Triune God, who is Father, Son, and Spirit—the beatific vision—and completed when our bodies are resurrected to a new heavens and a new earth, will not be achieved by evolution or by any natural process, but by the salvation won for us by Jesus Christ who is true God and true man.

Thomas Davenport, O.P.

❧ CHAPTER 31

In Defense of Thomistic Evolution

In his recent book, *Aquinas and Evolution* (Chartwell Press, 2017), my Dominican brother Fr. Michael Chaberek, O.P. has argued that Aquinas' thought cannot be reconciled with evolutionary biology without doing irreparable violence to the Thomistic synthesis. He is convinced that the Thomistic philosophical framework is incompatible with Darwinian evolution.

If Aquinas were alive today, how would he think through the process of evolution? Would he conclude that it is philosophically impossible?

It is clear that Aquinas did not know that organisms evolved. Like most, if not all, persons in Christendom during the thirteenth century, he believed on the authority of divine revelation that most of the organisms belonging to the natural kinds we see in the biological world were created directly by God and reproduced according to their own kind. It is striking, however, that he did acknowledge that at least one biological natural kind, the mule, could not have been directly created by God because it is the offspring of two other natural kinds, an ass and a mare, which God had to create first.[1] Nonetheless, Aquinas acknowledged that the creation of the mule could still be attributed to God because mules "existed previously in their causes."

In this concluding chapter, I would like to examine the five reasons/objections that Fr. Chaberek presents in the second chapter of his book for why Aquinas' teaching, in principle, necessarily excludes evolution.[2] I will provide rebuttals to illustrate the coherence of a Thomistic account of evolution. My basic thesis is that on Aquinas' own terms, it is still philosophically intelligible to claim that God created through evolution.

OBJECTION 1: *"According to theistic evolution, the lower (i.e., less perfect) cause can lead to the higher effect (i.e., more perfect). But in Aquinas' view, no being can convey more act than it possesses."*

RESPONSE: As I read Aquinas here, a more perfect being is something that has more capacities, more powers, than a less perfect substance. These novel capacities and powers constitute it as something that has "more act" than its less perfect counterparts. With this in mind, Fr. Chaberek's objection is a robust one based on a foundational metaphysical principle in the Aristotelian-Thomistic synthesis. The less perfect cannot make the more perfect. In everyday language, the principle is relatively straightforward: You cannot give what you do not have.

In response, let us consider a specific example from evolutionary biology: the evolution of snakes from lizards. The consensus among evolutionary biologists is that snakes evolved from lizards over a hundred-million-year period. To put it another way, from the perspective of evolutionary biology, a snake, for the most part, is a lizard that has lengthened its body, lost its limbs, fused its eyelids, and then made them transparent.

---

1. Cf. *ST* I, q. 73, a. 1, ad 3.
2. This chapter was first published online in *Public Discourse* on March 7, 2018. Available at https://www.thepublicdiscourse.com/2018/03/20975/. Last accessed on August 5, 2019.

Notably, evolutionary developmental biologists have already identified some of the genetic events that could have contributed to this evolutionary transformation. For example, there are data that reveal that mutations in the *Sonic Hedgehog (Shh)* gene enhancer can explain the loss of the rear limbs in the python. There are other data that reveal that mutations in the *Hox* genes can explain the transformation of the lizard's spinal column from lumbar vertebrae without ribs to thoracic vertebrae with ribs. (A snake has an extended thoracic region compared to that of a lizard.) Finally, we have also identified the molecular mechanism to increase the number of vertebrae in an animal spine. (A typical snake has over 300 vertebrae, while a typical lizard only has around sixty-five vertebrae. Humans have thirty-three vertebrae.)

For the sake of brevity, I summarize this evolutionary transformation that gave rise to the first snake using the following equation:

$$\text{lizard}_1 + \text{lizard}_2 \rightarrow \text{snake}$$

Here, we imagine mating pairs of lizards, where each lizard has a unique set of mutations, which, when put together into a single genome through mating, specify snakes with elongated bodies, no limbs, and fused transparent eyelids.

Fr. Chaberek's objection would apply here: How could two individual lizards, each a substance with lizard nature, interact to give rise to a snake, which is a substance with a snake nature? (Note that the snake nature has novel aspects like transparent fused eyelids that lizard natures lack.) Fr. Chaberek claims that they could not. In his view, therefore, evolutionary theory is ruled out on philosophical grounds.

Before I respond to Fr. Chaberek's objection to evolution, I think that it is important to point out that his philosophical objection is an objection not only to modern biology but to modern chemistry as well. Consider the following equation from basic high school chemistry:

$$\text{hydrogen} + \text{oxygen} \rightarrow \text{water}$$

Here, the substance of hydrogen interacts with the substance of oxygen to give rise to the substance of water.

Fr. Chaberek's objection applies here too. Philosophically, how can two unique substances of two different natural kinds, hydrogen and oxygen, give rise to another substance of another natural kind, water, with distinctive properties and chemical traits that they themselves do not have?

Clearly, I think that the claims of modern chemistry are true, and that hydrogen and oxygen can indeed interact to generate water. Fr. Chaberek, himself, I am sure, would agree. How then do we explain this chemical transformation *philosophically* from within the Thomistic tradition?

We have two options.

First, as Aquinas himself did, we could include the natural causality of other more perfect natural substances that mediate this and all the other chemical transformations that chemists study in our explanation of how chemistry works.

In his own day, Aquinas believed that the causality of the angels and of the celestial spheres or heavenly bodies—perfect substances thought to exist in the heavens by the best thinkers of his day—was involved in all the natural processes of change.[3] Since the angels and the celestial spheres are more perfect than any material thing, they would give to hydrogen and to oxygen the perfections that they lack so that together they can generate water.

---

3. cf. *De potentia*, q. 5, a. 8; *De veritate*, q. 5, aa. 8–9; *De operationibus occultis*.

Second, we could forsake natural causality entirely and include the supernatural causality of God in our philosophical explanation of everyday chemical transformations. Here God, who is the most perfect of all beings, would provide the perfections lacking to hydrogen and to oxygen so that with his help, they could together generate water.

Returning to evolutionary biology, I think that we can explain evolutionary transformations using the same philosophical strategies summarized above to explain chemical ones. We could invoke the causality of more perfect natural substances—in the absence of celestial spheres, we could speak about the causality of angels alone—or we could invoke the causality of God. Aquinas himself appeals to the power of the heavenly bodies to explain how fraternal twins of different sexes are created.[4] The esteemed twentieth-century Thomist philosopher, Jacques Maritain, chose the second option when he posited the existence of a "superelevating" motion of the First Cause working in tandem with the immanent activity of organisms to explain biological evolution.

Regardless of how one chooses to philosophically explain evolutionary transformations (and chemical transformations, for that matter!), these conceptual strategies suggest credible approaches to respond to Fr. Chaberek's first objection to the tenability of a Thomistic account of evolution. If hydrogen and oxygen can together produce water, then two individual lizards should, in principle, be able together to produce a snake.

OBJECTION 2: "*The second reason theistic evolution contradicts Aquinas' doctrine is that it presupposes that the nature (or the substantial form) of a living being can be changed into a different nature by an accidental change. This, however, is impossible in*

---

4. Cf. *De veritate*, q. 5, a. 8, ad 9.

Aquinas' view: accidental change can lead to only accidental changes, whereas a change of nature requires substantial change."

RESPONSE: To understand this objection, we first need to review Aquinas' account of change. Like Aristotle, Aquinas acknowledged the commonsense view that there are two kinds of change in the world. The first type, called substantial change, involves radical change that transforms an individual of one kind into an individual of another kind. A human being becoming a corpse is an example of this kind of change. The second type, called accidental change, involves superficial change that transforms an individual without changing his fundamental identity. A fat human being becoming a thin human being after an all-fruit diet is an example of this second kind of change.

To explain these two kinds of changes, Aquinas, again like Aristotle, posited the existence of two constituents of individual things. First there is form, which is that constituent that explains what something is, and another correlative constituent called prime matter that explains how that something can perdure through change. Every individual thing is made up of prime matter and a substantial form that informs it. For material things, one cannot have matter without form, nor form without matter.

In Aquinas' view, therefore, substantial changes involve a transformation of substantial form in which prime matter remains the same. Thus, when a human being dies, his prime matter, once informed by his human substantial form, is now informed by the substantial forms of the elements, primarily carbon and hydrogen.

Given Aquinas' account of change, it is clear that every evolutionary transformation, like the lizard-to-snake transformation discussed above, must necessarily involve a change in substantial form. It requires a substantial change.

Fr. Chaberek objects that theistic evolution assumes "the nature (or the substantial form) of a living being can be changed into a different nature by an accidental change . . . This, however, is impossible in Aquinas' view: accidental change can lead to only accidental changes." I agree and disagree.

Fr. Chaberek is correct when he says that accidental changes cannot lead to substantial changes *directly*. However, I do think that accidental changes can lead to substantial changes *indirectly*. They do this by changing prime matter's disposition to substantial form.

Consider the directionality and specificity of change. When a human being dies, he becomes a corpse, and not a kaleidoscope of butterflies. When a log burns in a fire, it becomes ashes and not a puddle of water. To explain this specificity and directionality of change, Aquinas appealed to the disposition of prime matter.

In Aquinas' view, a substantial form limits the disposition of prime matter such that logs become ashes and not water, because the matter of logs is predisposed, by the substantial form of wood, to the substantial forms of the elements in ashes rather than to the substantial form of water. In the case of human death, a human being becomes a corpse rather than butterflies, because the human substantial form predisposes the matter in the human body to the forms of carbon and hydrogen and not to the substantial forms of butterflies.

Importantly, Aquinas also appeals to the disposition of prime matter to explain how different individuals of the same natural kind can possess natural traits to different degrees. Thus, he will explain that a person with exceptional intelligence is an individual whose matter is better disposed to his human form than is typical of human beings.[5]

---

5. Cf. In II *de Anima*, 19.

Returning to Fr. Chaberek's objection, I propose that accidental changes can lead to substantial changes *indirectly*, because accidental changes can change the predisposition of matter to a substantial form. Aquinas wrote: "Every substantial form requires a proper disposition in its matter, a disposition without which it cannot exist. This is why alteration is the way to generation and to corruption" (*De mixtione elementorum*).

Recall my example of an accidental change where a fat man becomes a thin man after an all-fruit diet. If we alter this example slightly, we get the following scenario: A fat man becomes a thin man who becomes a dead man after a month-long hunger strike. Biologically, we would say that dehydration and starvation led to multiple organ failure and the death of the hunger striker, but how do we explain this process of change *philosophically*? We would have to say that the accidental changes that gradually led to the decrease in the man's overall mass also altered the predisposition of his matter such that a substantial change, i.e., his death, inevitably occurred.

In the same way, I propose that genetic mutations can also affect the predisposition of matter. Recalling Aquinas' explanation for the superior intelligence of certain persons, for example, I think that an individual with Down syndrome is not as intelligent as the rest of the human population because his extra twenty-first chromosome alters the predisposition of his matter so that it is not as predisposed to his substantial form as it is in his peers.

Therefore, contrary to Fr. Chaberek's objection, in my view, accidental changes can lead to substantial changes *indirectly*. They do this by changing matter's disposition to substantial form.

OBJECTION 3: "*The third reason is that theistic evolution presupposes that one nature can be the cause of another nature. In contrast, according to Aquinas no 'perfect thing' produces its own nature, but only participates in the nature that it inherits… Thus, a man cannot be a cause of mankind, a dog cannot cause dog's nature, a cat, a cat's nature, and so on. But if a being is not a cause of its own nature, much less can it produce a different nature (ScG III.69 c)."*

RESPONSE: Fr. Chaberek is correct. According to Aquinas, a being cannot cause its own nature, much less produce a different nature. However, this philosophical objection does not rule out evolutionary transformations.

Taking what we have already said to respond to Fr. Chaberek's first two objections to Thomistic evolution, we can now provide the following philosophical account for evolutionary transformation that does not require that one nature be the cause of another nature.

Genetic mutations lead both to accidental changes and to changes in the disposition of matter. For example, take the following hypothetical scenario: a mutation in the *Sonic Hedgehog* (*Shh*) gene enhancer could lead to four-limbed lizards giving rise to lizard offspring that lack rear limbs. This would be accidental change. However, philosophically speaking, this *Shh* mutation would also alter the disposition of the two-legged lizard's matter such that it is now predisposed to the novel substantial form that specifies two-legged lizards as a distinct natural kind. If this mutant two-legged lizard were to mate with another mutant lizard bearing the same *Shh* genetic mutation, then they would generate lizard offspring that permanently lacked rear limbs. This would be substantial change. Strikingly, stable two-legged lizard species exist today that have lost either both of their front limbs or both of their hind limbs. There are even species of legless lizards that look like snakes but differ from snakes in other ways!

Here the generation of the new natural kind of two-legged lizards would not be attributed to the causality of the four-legged lizard—Fr. Chaberek's concern in his objection—but to the causality of the First Cause working with and through creaturely causality in the process we call evolution.

An objector may point out that most genetic mutations are deleterious. They alter the disposition of matter such that progeny that bear these mutations are sickly and die. This is true. However, not all mutations are necessarily deleterious. As we saw above with the legless lizards, the loss of lizard limbs is not inherently lethal. Though we will never really know the reasons for the survival advantage that legless lizards have over their legged counterparts in a particular environmental scenario, it should not be too hard to see how mutant legless lizards would be able to burrow better into the ground to avoid predation or to chase prey than their four-legged parents.

OBJECTION 4: "*The fourth reason is that according to Aquinas (following Aristotle) every composite thing has four causes: final, formal, efficient, and material. In theistic evolution, the efficient cause of species is the power of generation only guided by the final cause, who is God. Thus, even though theistic evolution involves final causality, it reduces 'down' efficient causation to accidental changes in matter and the operations of nature (medieval 'movements of heavens'). And, according to Aquinas, this kind of cause cannot generate new natures (as was shown in the third problem) . . . And since it is formal cause (not matter), which produces the substantial form, theistic evolution lacks the formal cause, which is reduced 'up' to the final cause alone. Therefore, theistic evolution lacks both formal cause and efficient cause.*"

RESPONSE: Fr. Chaberek is correct. The scientific revolution that replaced Aristotle's substantial biology with Descartes', Boyle's, and Locke's mechanical biology in the seventeenth century set out to discard formal and final causality. However, as I have written elsewhere at much greater length, today's genomic revolution has challenged biologists to recover a holistic account of the organism that embraces all four Aristotelian causes. I will only briefly summarize this view, which I call the systems hylo-morphic view, here.

Today, systems biologists view the organism as a dynamic network of molecular interactions over time. Here, the human animal is a substantial being, a dynamic network of macromol-ecules now existing not as independent molecules per se but as different virtual parts of the one human organism. This spe-cies-specific network, which is distributed in three-dimensional space, and which is able to interact over time in the robust, self-organizing process that we call human development, in my view, is a manifestation of the human being's substantial form. The correlative metaphysical principle that is actualized by that substantial form would be its matter.

Human embryos become human adults and puppies become dogs because of their networks of molecular interactions. Thus, the systems account described above not only emphasizes the holism (and thus the formal cause of the organism) but also the end-directedness of its development (and thus its final cause).

Fr. Chaberek is correct: classical evolutionary theory does not consider formal and final causes. However, he does not seem to realize that this is not a necessary aspect of the theory. There are already accounts of evolution that acknowledge that evolv-ing organisms are end-directed, holistic, and dynamic systems. These dynamic systems accounts reveal that evolutionary theory can properly take efficient, material, formal, and final causality into consideration.

OBJECTION 5: "*The fifth reason is that, according to Aquinas, God wanted different degrees of perfection in nature. This happens among different species, as well as within one organism—among its organs... Hence according to Aquinas, things less perfect and more perfect exist for the sake of the greatest perfection of the whole mate-rial world. This order is intended by God... And this is contrary to*

the evolutionary view of nature, in which each part requires contin-
ual change toward greater perfection in 'struggle for life' and 'survival
of the fittest."

In my view and the view of other contemporary proponents
of theistic evolution, God created through evolution. Why did
God create this way? For at least two reasons.

First, as we noted in an earlier chapter, for the Catholic tradition, the answer to the purpose-of-creation question is clear: God
chose to create because he wanted to manifest and communicate
his glory. The *Catechism of the Catholic Church*, the authoritative
summary of Catholic doctrine, proclaims that "Scripture and
Tradition never cease to teach and celebrate this fundamental
truth: 'The world was made for the glory of God.'"

How does God communicate his glory to his creatures?
According to Aquinas, God communicates his glory to his creatures by giving them a participation in his existence. However,
he also explains that God shares his perfections with his creatures
by inviting them to participate in his causality, which manifests
itself in his governance of his creation:

> But since things which are governed should be brought
> to perfection by government, this government will be so
> much the better to the degree that the things governed
> are brought to perfection. Now, it is a greater perfection
> for a thing to be good in itself and also the cause of good-
> ness in others, than only to be good in itself. Therefore,
> God so governs things, that he makes some of them to
> be causes of others in government, like a master, who not
> only imparts knowledge to his pupils, but gives also the
> faculty of teaching others.[6]

---

6. *ST* I, q. 103, a. 6

To put it another way, according to Aquinas, it is a greater perfection, and therefore is more fitting, for God to share his causality with his creatures, making them authentic causes that can cause by their own natures along with God, than for God to remain the sole cause acting within the universe.

By creating through evolution, God is able to invite his creatures to work with him to generate the novelty and diversity of life. As Aquinas noted, this is a greater perfection than if he had chosen to create life on his own via special creation.

Second, according to Aquinas, God also created the diversity of creatures because no single creature can adequately reflect the perfection of God:

> We must say that the distinction and multitude of things come from the intention of the first agent, who is God. For he brought things into being in order that his goodness might be communicated to creatures, and be represented by them; and because his goodness could not be adequately represented by one creature alone, he produced many and diverse creatures, that what was wanting to one in the representation of the divine goodness might be supplied by another. For goodness, which in God is simple and uniform, in creatures is manifold and divided and hence the whole universe together participates in the divine goodness more perfectly, and represents it better than any single creature whatever.[7]

Therefore, in my view, it is also fitting that God worked via evolution rather than via special creation, because in doing so he was able to produce more species to reflect his glory. Four billion

---

7.   *ST* I, q. 47, a. 1.

species created over a three-billion-year period is far more than the eight million extant species today. In fact, it would have been ecologically impossible for all four billion species to coexist on our planet, because there is only a limited number of ecological niches on the planet at any given moment in time.

If they had been created together, for instance, the large carnivorous dinosaur, *Tyrannosaurus rex*, would have wiped out the Asian elephant, *Elephas maximus*. However, with evolution—and not with special creation—these natural kinds were able to exist at separate moments in history to uniquely manifest the glory of their Creator.

To sum up, why did God choose to work via an evolutionary process rather than will a special creation? Because it better reveals his glory and his power. Because it reveals better that he is God.

Fr. Chaberek claims that an evolutionary view of creation would undermine God's intent to create a world with varying degrees of perfection because each natural kind "requires continual change toward greater perfection in 'struggle for life' and 'survival of the fittest.'" But I do not think that an evolutionary biologist today would claim that evolution is a movement toward greater perfection. The dean of evolutionary biologists, the late Stephen Jay Gould, certainly denied this explicitly in his book, *Full House*. Instead, he claimed that evolution is a movement toward a greater diversity of life. And as I noted above, this would not be contrary to God's intent of creating a material world that is ordered towards his glory.

The increase in the number and kinds of organisms over evolutionary history better reflects the infinite ways in which finite creatures could reflect the infinite beauty of God. Evolution does not undermine divine providence.

Nicanor Pier Giorgio Austriaco, O.P.

❧ About the Authors

FR. NICANOR PIER GIORGIO AUSTRIACO, O.P., currently serves as Professor of Biology and of Theology and as Director of ThomisticEvolution.org at Providence College. He is also a Research Fellow of the Center for Religious Studies and Ethics at the University of Santo Tomas in the Philippines. Fr. Austriaco received his Ph.D. in Biology from M.I.T. and his Doctorate in Sacred Theology (S.T.D.) in Moral Theology from the University of Fribourg in Switzerland. His NIH-funded laboratory at Providence College (www.austriacolab.com) is investigating the genetics of programmed cell death using the yeasts *Saccharomyces cerevisiae* and *Candida albicans* as model organisms. His first book, *Biomedicine and Beatitude: An Introduction to Catholic Bioethics*, was published by the Catholic University of America Press. It was recognized as a 2012 Choice outstanding academic title by the Association of College and Research Libraries.

FR. JAMES BRENT, O.P., earned his Ph.D. in philosophy from St. Louis University and his License in Sacred Theology (S.T.L.) from the Pontifical Faculty of the Immaculate Conception at the Dominican House of Studies in Washington, DC. He is currently Assistant Professor of Philosophy at the same Pontifical Faculty in Washington, DC.

FR. THOMAS DAVENPORT, O.P., is Assistant Professor of Physics at Providence College. He received his Ph.D. in physics from Stanford University and his License in Philosophy (Ph.L.) from the Catholic University of America.

FR. JOHN BAPTIST KU, O.P., earned his Doctorate in Sacred Theology (S.T.D.) from the University of Fribourg. His dissertation was awarded the 2010 St. Thomas Aquinas Dissertation Prize by the Aquinas Center for Theological Renewal at Ave Maria University in Florida, and was published as a monograph, *God the Father in the Theology of St. Thomas Aquinas,* by Peter Lang International Academic Publishers. Fr. Ku is currently an Associate Professor of Theology at the Dominican House of Studies at the Pontifical Faculty of the Immaculate Conception in Washington, DC.

Cluny Media

Designed by Fiona Cecile Clarke, the Cluny Media *logo depicts a monk at work in the scriptorium, with a cat sitting at his feet.*

The monk represents our mission to emulate the invaluable contributions of the monks of Cluny in preserving the libraries of the West, our strivings to know and love the truth.

The cat at the monk's feet is Pangur Bán, from the eponymous Irish poem of the 9th century. The anonymous poet compares his scholarly pursuit of truth with the cat's happy hunting of mice. The depiction of Pangur Bán is an homage to the work of the monks of Irish monasteries and a sign of the joy we at Cluny take in our trade.

"Messe ocus Pangur Bán,
cechtar nathar fria saindan:
bíth a menmasam fri seilgg,
mu memna céin im saincheirdd."

Printed in the USA
CPSIA information can be obtained
at www.ICGtesting.com
LVHW051237191223
766490LV00030B/765/J